# Sequential Experiments with Primes

Mihai Caragiu

# Sequential Experiments
# with Primes

 Springer

Mihai Caragiu
Department of Mathematics and Statistics
Ohio Northern University
Ada, OH
USA

ISBN 978-3-319-85995-8          ISBN 978-3-319-56762-4    (eBook)
DOI 10.1007/978-3-319-56762-4

Printed on acid-free paper

This Springer imprint is published by Springer Nature
The registered company is Springer International Publishing AG
The registered company address is: Gewerbestrasse 11, 6330 Cham, Switzerland

# Preface

## Why This Book?

This book is actually about the mathematical life and destiny of mathematics faculty and their talented students in small undergraduate colleges (not necessarily the elite ones, which are different) who wish to obtain a glimpse into the ethereal world of "higher mathematics." How can this be done, in spite of daily pressures such as high teaching and service loads for faculty and the heterogeneous and career-oriented curricular schedules for students? How can these students learn to see the value of higher mathematics? These things need to be figured out for an education that will prepare students for lifetime of learning.

My experience as a mathematics teacher at Ohio Northern University has led me to an answer that I would like to share by means of this book with other faculty members and talented students at small undergraduate colleges: it is about experimenting with elementary number theory (prime numbers and related functions) and witnessing the amazing behavior of special integer sequences. Through elementary means we managed in six years to get from scribbling a recurrence formula on a piece of paper to being spoken of at a 2012 international conference on Fibonacci numbers. For a small undergraduate college, that was a big deal.

Ada, OH, USA                                                            Mihai Caragiu
January 2017

# Acknowledgements

So… students first! I would like to thank for inspiration, enthusiasm, and computer expertise the wonderful generation of students who participated in our "GPF Sequences" projects, got to experience extracurricular excursions in algebra and number theory in cooperation with the author, or participated in the "ONU-Solve" problem club: Andrew J. Homan (graduated in 2008 with degrees in mathematics and philosophy), Greg T. Back (graduated in 2010 with degrees in computer engineering and mathematics), Justin Gieseler (graduated in 2010 with a major in computer engineering and a minor in mathematics), John T. Holodnak (graduated in 2010 with a degree in mathematics), Ashley M. Risch (graduated in 2011 with a degree in mathematics), Lauren T. Sutherland (graduated in 2011 with degrees in electrical engineering and mathematics), Donald J. Pleshinger (graduated in 2014 with degrees in physics and applied mathematics), Jonathan C. Schroeder (graduated in 2014 with degrees in computer engineering and mathematics), Thomas E. Steinberger (graduated in 2014 with degrees in physics and mathematics), Thomas J. Gresavage (graduated in 2014 with degrees in mechanical engineering, German, and applied mathematics), Lisa A. Schekelhoff (graduated in 2015 with a triple degree in pharmacy, biology, and mathematics), Michelle E. Haver (currently a mathematics major, graduating in 2017), Matthew Golden (currently a physics and mathematics major, graduating in 2017), and last but not least, Paul A. Vicol, of Simon Fraser University (graduate student).

Many thanks go to faculty participants in the "GPF" and related projects: Jaki Chowdhury, of Ohio Northern University; Ronald A. Johns, of Ohio Northern University; and Alexandru Zaharescu, of the University of Illinois at Urbana Champaign. Also, many thanks go to Cristian Cobeli, of the Institute of Mathematics in Bucharest, whose profound work on interesting sequences inspired me and helped shift my attention to general recurrences with Conway's "subprime function" and related non-associative structures at a moment when I was becoming perhaps too preoccupied with the greatest prime factor sequences at the expense of other ideas.

As a former student myself, I owe words of thanks in temporal order, to Serban Basarab (1940–2014), of the Institute of Mathematics in Bucharest, for guiding my first serious steps at the interface between finite fields and logic, and to Leonid Vaserstein, of Pennsylvania State University, for being a great, encouraging, and inspiring advisor during my life as a Penn State grad student and who thus helped define my mathematical life.

Last but not least I would like to give thanks for constant professional support to Springer editorial director Marc Strauss, assistant editor Dimana Tzvetkova, and the book production team including Shobana Ramamurthy, Karthik Raj Selvaraj, and Swetha Puli.

# Contents

# Chapter 1
# Introduction

## 1.1 Low-Budget Space Travel

A frequent key phrase often heard in undergraduate colleges is "high-impact learning." Indeed, the Association of American Colleges and Universities (AAC&U) issued a list of ten "high impact educational practices" (Kuh 2008) that would arguably boost student success, out of which two are of particular interest to this work: undergraduate research, and capstone courses and projects.

The present book will consider ways of boosting success in mathematics for students at small undergraduate colleges. It will consist of a grab bag of new (to the best knowledge of the author) mathematical ideas and problems (most of them open problems, with some of them proved) involving prime numbers and related sequences that the author hopes will boost the enthusiasm for exploration in pure mathematics of both students and faculty at small undergraduate colleges.

In the opinion of the author, disseminating and advertising the core mathematical ideas (those routinely grouped under the "pure mathematics" label) face a variety of objective challenges in today's context that need to be acknowledged. The rapid growth of business, investment, consulting, and insurance companies (needless to say, well funded and offering attractive salaries) creates a high demand for graduates "good with numbers," implicitly generating changes in the expectations of mathematics (and statistics) majors (for example, the rapid growth of the insurance industry tilts the demand scale towards bachelor's degrees in mathematics or statistics with an actuarial science concentration). Relatively few students today are eager to engage in "hopelessly pure" areas such as number theory, geometry, combinatorics, or analysis. For incoming freshmen, "being good with numbers" in the context of a possible career after graduation is, most of the time, associated with careers in accounting, business, and management: being good with numbers is only marginally associated with, say, prime numbers.

© Springer International Publishing AG 2017
M. Caragiu, *Sequential Experiments with Primes*,
DOI 10.1007/978-3-319-56762-4_1

That is why mathematics faculty engaged in undergraduate education need to find ways for a "new beginning" that can jump-start the enthusiasm for that amazing body of knowledge commonly known as "pure mathematics."

For applied-minded students who are pursuing a degree in computer engineering, one can emphasize, for example, the tremendous success of number theory and related cryptosystems in the "hot areas" (by the standards of today's society) of information security—an activity that may be construed as a significant "outreach" on behalf of the "queen of mathematics" (number theory, as seen by Carl Friedrich Gauss).

For those students with interests in the fundamental physical sciences, we can emphasize the important role of geometry, topology, and algebra in the areas of physics in which the "big questions" reside (quantum theory and gravitation/cosmology); for example, all three 2016 Nobel laureates in physics used topological ideas to guide their explorations involving new phases of matter and topological phase transitions (Nobelprize.org 2016).

For students interested in applications of statistics, the mathematics teacher can suggest that they may wish to focus their talent on foundational issues of probability theory and random processes (after all, modern probability theory, a pure mathematical theory, was founded by Kolmogorov in the 1930s), and possibly—to offer just two examples—to follow up by applying these fundamental results to the study of the immutable reality of prime numbers [e.g., the special Poissonian character of the distribution of primes (Gallagher 1976)] or to mathematical economics (if they have seen the movie "A Beautiful Mind," that would definitely be a plus).

In any case, this is not about trying to get students to "switch" their major or concentration of study to pure mathematics. There are students with genuine interest in business, analytics, applied statistics, actuarial science, physics, engineering, environmental science, chemistry, etc., which is great. At the same time, however, a passionate mathematics teacher should always try to provide students with opportunities to witness firsthand some aspects of the beauty and the depth of mathematics. Getting them to experience pure mathematics would be enriching, and for those students having, in fact, a "pure" mathematics sensibility, that would be a genuine moment of self-discovery.

Everything sounds nice… if it weren't so difficult. The typical small undergraduate college is not in the "top 20" elite, does not have a huge financial endowment, and is faced with many challenges, especially when it comes to admissions. Getting any students at all to major in mathematics is at times difficult. Many incoming students are confused, unsure about the path they should take, and the subtleties of upper-level mathematics courses make them uncomfortable.

The dissemination of the beauty of the pure mathematics among students is generally harmed by a variety of factors. In the opinion of the author, the leading such factor, especially when it comes to mathematics or statistics majors, is downplaying—or simply not even emphasizing enough—the importance of extracurricular activities such as group projects involving solving problems proposed in various mathematics journals or discussing various important mathematical ideas, participating in summer Research Experiences for Undergraduates, presenting at mathematical conferences, or simply "writing" mathematics. Neglecting such extracurricular activities arguably limits a student's creativity and the desire to pursue graduate work in mathematics.

Also, there are the suggestions and promises of generally well-paid desk jobs for new graduates (in business, banking, management, insurance, etc.) after an accordingly narrow course of study that avoids the deeper end of the mathematical content. This makes courses such as abstract algebra, real analysis, geometry, topology, combinatorics, advanced random processes, and number theory, appear unnecessary, burdensome, uninteresting, irrelevant, and unnecessarily cumbersome. Of course, the fact that in the "age of media streams" the average human attention span dropped from 12 s (in 2000) to 8 s (in 2015) (McSpadden 2015) does not really help (of course, the author does not make any claim of not being part of this trend).

So, the real question is, what can we do (as college teachers), in this particular context, to increase students' exposure to mathematical ideas that might awaken their "researcher within" and subsequently send them on a path of discovery in mathematics?

I don't think that asking students to read renowned classic mathematical monographs (or voluminous collections of classical articles) is a feasible solution, especially today, and especially at the "typical" small undergraduate college. Instead, we should exploit one particularity of the current age that has already proved time and again to be beneficial to the research mathematician: *the use of computers.*

The approach taken in this work is that if the mathematics professor manages to communicate the spirit of "experimental mathematics" to an interested student, indicating a problem that is elementary and easy to formulate (but generally hard to solve), then the computing environment becomes an extremely beneficial "instrument of dialogue" between the professor and the student. Mathematical reality can be investigated much as physical reality is investigated in particle physics: at higher and higher energies, many interesting and unexpected phenomena and particles are generated. In the same way, using (generally simple) programming and computer algebra systems by "mathematical experimentalists" (student and teacher) can generate interesting new mathematical ideas and conjectures.

In addition, when some of these conjectures (suggested as a result of computational analysis) can be proved through an approach that is reasonable for an undergraduate mathematics major (generally elementary, albeit intricate at places), the satisfaction is so much the greater. In the author's educational experience, students who participate in the process of discovering new and interesting mathematical knowledge (about prime numbers and related sequences in the case of the present work) through computation, to say nothing about some of it being actually proved, will have a first-hand, significant, high-impact learning experience that will change their views on pure mathematics for the better.

At the same time, the instructor can say, figuratively speaking, that a "low-cost space travel" experience has been made available to the student. "Low-cost" because it involves elementary ideas that are simple to formulate and then the use of a computing environment requiring generally simple programing and a convenient user interface. "Space travel" because the conclusions of the experiment are beautiful, interesting, and represent new mathematical truth, with some parts of it proved, while most other parts await further exploration.

This book is meant to inspire the pure-mathematics-minded educator in a typical small undergraduate college by offering a variety of elementary number theory insights involving sequences essentially built from prime numbers and associated number-theoretic functions, together with related conjectures and proofs. It is not meant to be "complete," nor can it ever be so, since one can imagine new number-theoretic functions and use appropriate programming to investigate other topics, in collaboration with students looking for a significant high-impact learning experience in mathematics.

Our life as college mathematics teachers seeking significant research experiences to offer our students is not exactly easy. Indeed, it is not easy to be in a "research-intensive" mode when the regular curriculum teaching load is 12 credit hours per semester. Yet I hope that this computational/experimental approach will help to make it easier.

The basic requirements for a typical participating student are an interest in mathematical discovery, an eagerness to face new ideas, an ability to read and do some basic proofs in elementary number theory, and a willingness to use computers to test the hypotheses that appear along the way.

The eagerness to face new ideas should be reflected in a passionate desire to test new results and discoveries using computers. Just to give an example, while Michelle Haver, one of our undergraduate students and a "pure" mathematics major, was considering a topic of presentation to the Ohio MAA "Centennial Meeting" that took place on April 8–9, 2016, at Ohio Northern University, we discovered the celebrated "prime conspiracy" result of Kannan Soundararajan and Robert Lemke Oliver (Klarreich 2016). Since we liked the topic and didn't have much time left before the upcoming MAA conference, we decided to try a sort of "Monte Carlo" simulation to study the interesting "self-avoidance" phenomenon between the congruence classes of pairs of consecutive primes discussed in the amazing paper (Lemke Oliver and Soundararajan 2016), to the effect that, for example, if a prime ends in 1, the following prime is less likely to end in 1. Thus we decided to verify the "self-avoidance" conjecture using MAPLE by randomly selecting pairs of consecutive large primes up to 12 digits and analyzing the set of corresponding last-digit pairs. We wanted the output to be in the form of a histogram, because visualizing is believing. The four groups of four bins reflect the distribution of last-digit pairs after ten thousand random trials. In the first group, we see the counts for 11, 13, 17, and 19, in the second group the counts for 31, 33, 37, and 39, etc. The simple instruction line in MAPLE was as follows:

```
> for i from 1 to 10000 by 1 do p:=nextprime(rand(10^12)()):
q:=nextprime(p): L[i]:=10*(p mod 10)+(q mod 10): end do:
M:=[seq(L[i],i=1..10000)]: Histogram(M, frequencyscale =
absolute, discrete=true);
```

To our satisfaction, we noticed that most such computer experiments display the anticipated "self-avoidance" pattern. A typical such histogram, displayed below, is result of a Monte Carlo simulation involving 10,000 random choices for primes with up to 12 digits. In each group of bins, the "identical parings" (11 in the first group of four, 33 in the second group, 77 in the third group, and 99 in the fourth group) appear with the least count.

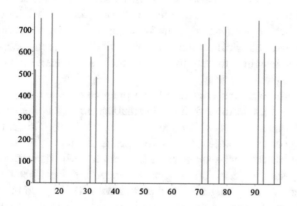

It is such experiments that empower students and help guide their path toward advanced study at the graduate level. There, where things get much more "serious" and the proofs harder, the mathematics student is helped by the enthusiasm gained during his or her undergraduate studies. Again and again, I would like to emphasize the importance of visualizing interesting and important number-theoretic patterns as an "enthusiasm boost" for undergraduate students. Needless to say, Michelle's talk at the 2016 Ohio MAA meeting (Haver 2016) elicited a healthy dose of interest from the audience, and also that this kind of student success (as instances of high-impact learning) marks a bright spot in the life of a mathematics teacher at an undergraduate college.

To summarize, this book is intended to be a "good read" on a series of new applications of elementary number theory with an experimental/computational emphasis involving sequences, primes, and special arithmetic functions (with an emphasis on the greatest prime factor function). Its goal is to enable students and faculty at small undergraduate colleges engage and pursue new ideas conducive to high-impact learning in a beautiful field.

This book is *not* meant to be a monograph ion computational and algorithmic number theory. First of all, there are already many excellent such works, in which foundations with a wealth of examples are addressed at length; see, for example,

Bach and Shallit (1996), Borwein (2002), Bressoud and Wagon (2000), Buhler and Stevenhagen (2008), Das (2013), von zur Gathen and Gerhard (2003), Shoup (2009) to cite just a few. Secondly, the author does not wish to claim any significant authority in matters of programming: without significant computer science training, I had to go pretty much by myself into matters of coding, receiving invaluable computing tips and inspiration from an amazing group of "virtual teachers" that are online for everybody's benefit: MATLAB MathWorks, MAPLE Help (or, online, MAPLE Primes), and the On-Line Encyclopedia of Integer Sequences (OEIS), which I tried to complement, over many years, by many hours of trial and error generating homemade number-theoretic functions for the benefit of both myself and some passionate students.

This book is about a set of special experiments with primes, selected arithmetic functions, and integer sequences, of a flavor echoing the "$3x + 1$" problem. Before getting to our "weird" greatest prime factor sequences, we provide (in Section 1.2) a quick overview of classical results in number theory and recurrences that provide background and insight into helpful "trivia" for the sequential experiments to be conducted subsequently.

In many of the examples considered, the practice of experimental mathematics reflects a certain attitude toward the mathematical objects being investigated (for example, the prime numbers, which play a special role here). This attitude reflects a tendency to incorporate such objects in special contexts (or surroundings, or "worlds," such as recurrent sequences of integers and sets or random structures).

In studying such objects, one begins by recording their behavior in various contexts. Common sense suggests that their behavior will depend on their "intrinsic nature" on the one hand, and on the specific particularities of the environment in which they are allowed to evolve. If such behavior reveals particularly stable patterns, one then searches for proofs or formulates conjectures. One can also change the contexts in the hope that repeated analysis of the objects' behavior will reveal some meaningful truths about their intrinsic nature.

Computational and experimental methods, alongside exquisite number-theoretic proofs that lie at the interface between "easy" (or so they appear) and "wow" are excellent tools and supplements for an undergraduate classroom in number theory and its applications.

Since our objects of study are the prime numbers, which quite often can be incorporated into "sequential" contexts in which mathematical experiments unfold, this book could well of have been called *Sequential Experiments with Prime Numbers*.

## 1.2   A Topical Overview

In Chapter 2 we present some background information and a series of themes and ideas that periodically surfaced in our activities with undergraduate students engaged in senior capstone projects in number theory, or just students wishing to do research in number theory. We outline a few projects involving computational and experimental mathematics, with an emphasis on prime numbers, primitive roots, but also applied-leaning topics in number theory inspired by the search for random structures and traffic flow theory. As background for the material coming next, we discuss computational exploration of periodicity of recurrent sequences, accompanied by examples of cases in which we used MAPLE or MATLAB to attack specific problems. The emphasis, though, is on the quest for discovery of new and interesting mathematical content.

Greatest prime factor sequences are explored in detail in Chapter 3. This is arguably the central theme of this book. The study of such sequences originated in an undergraduate institution (Lisa Scheckelhoff, Ohio Northern University) in the fall quarter of the academic year 2005–2006. Repeated computer experiments with different values of the parameters $a, b$ suggested the earliest version of the "GPF conjecture": all prime sequences satisfying a recurrence relation of the form $x_n = P(ax_{n-1} + b)$ (first-order greatest prime factor sequences or "GPF sequences") are ultimately periodic. This is discussed in Section 3.1. In said Ms. Scheckelhoff's capstone project, the GPF conjecture was proved for the special case in which $a$ divides $b$. We will see that multiple limit cycles are possible, while a general cycle structure result remains elusive.

Higher-order GPF sequences are discussed in Section 3.2. This originated with the 2009/2010 research projects of another student (Greg Back, a double major in computer engineering and applied mathematics), who did his senior research involving the greatest prime factor function. We considered GPF sequences of arbitrary order $d$, that is, sequences of primes $\{x_n\}_n$ satisfying a recurrence relation of the form

$$x_n = P(c_1 x_{n-1} + c_2 x_{n-2} + c_3 x_{n-3} + \cdots + c_d x_{n-d} + c_0).$$

Again, extensive computational evidence suggested that every such sequence is ultimately periodic. That was all very well, except that we felt the need of a completed proof of at least one partial result that backed the GPF conjecture in this general form. The inspiration came from Fibonacci numbers. The GPF analogue of the Fibonacci recurrence is sequences $\{x_n\}_n$ satisfying the "GPF-Fibonacci" recurrence $x_n = P(x_{n-1} + x_{n-2})$. In a paper published in the *Fibonacci Quarterly* in 2010, Greg Back and I managed to prove that for every choice of "initial conditions" $x_0, x_1$, every sequence of primes satisfying the above GPF-Fibonacci recurrence ultimately enters the same, unique, limit cycle $[7, 3, 5, 2]$. In the same paper, a computational analysis of the GPF-tribonacci sequences, namely those satisfying the recurrence $x_n = P(x_{n-1} + x_{n-2} + x_{n-3})$, was performed. Fairly extensive computational evidence suggests that the GPF-tribonacci sequences are ultimately periodic with (at least) four possible limit cycles, of lengths 100, 212, 28, and 6. There is an interesting lack of symmetry reflected in the fact that for a random choice of the "seed" $x_0, x_1, x_2$, the odds of entering one or the other of these four cycles are wildly uneven, the larger cycles collecting more than 98% of the GPF-tribonacci sequences as defined by the seed.

Multidimensional greatest prime factor sequences (MGPF) are discussed in Section 3.3. Again, these originated with another senior capstone project completed in 2011 by Lauren Sutherland (again, a double major in computer engineering and mathematics).

In Section 3.4, we discuss interesting results involving the set of the first four primes $\{2,3,5,7\}$. We already know that there is something special about the four primes, 2, 3, 5, and 7: in the order 7, 3, 5, 2 (up to a cyclic permutation) they constitute the limit cycle of all GPF-Fibonacci sequences. But there is more to it. The set $\{2,3,5,7\}$ is closed under the "greatest prime factor of the sum" binary operation defined as $x * y = P(x+y)$. Now, if we endow the set $\Pi$ of all primes with the binary operation $*$, then $(\Pi, *)$ is a nonassociative "magma." Every set with one element is a trivial substructure (or "submagma") of $(\Pi, *)$, but $K_7 := \{2,3,5,7\}$ is the unique minimal substructure of $(\Pi, *)$ in the sense that $\{2,3,5,7\}$ is included in every nontrivial submagma of $(\Pi, *)$.

The next two sections are a joint work with Alexandru Zaharescu, of the University of Illinois, and Jaki Chowdhury (M. Zaki), of Ohio Northern.

Infinite-dimensional analogues are discussed in Section 3.5. These are sequences satisfying a recurrence of the form

$$x_n = P(x_{n-1} + x_{n-2} + x_{n-3} + \cdots + x_0).$$

The amazing fact is that these sequences can be described by a simple formula in terms of the sequence of primes $\{p_n\}_{n \geq 1}$. Indeed, for every sufficiently large integer $n$, the prime $p_n$ has $p_{n+1} - p_{n-1}$ consecutive occurrences in $\{x_n\}_n$. Moreover, using a certain upper bound for the gap between consecutive primes, it was proved (Caragiu et al. 2012) that the asymptotic behavior of such an infinite-order GPF sequence is given by the following remarkable formula:

$$x_n = n/2 + O(n^{0.525}).$$

In Section 3.6, we present interesting analogues of the iterative Ducci game, which incorporates the greatest prime factor function. The classical Ducci game starts with a vector of integers $(x_0, x_1, x_2 \ldots, x_{n-1})$ and keeps updating it as $(|x_0 - x_1|, |x_1 - x_2|, |x_2 - x_3| \ldots, |x_{n-1} - x_0|)$, thus generating a sequence of integer vectors. It is proved that every such Ducci iteration enters a cycle, and that every vector in the limit cycles has components 0 or C, where C is a fixed positive integer throughout the cycle. Our GPF analogue involves sequences of primes $(x_0, x_1, x_2 \ldots, x_{n-1})$ that are updated, using the greatest prime factor of the sum instead of the absolute value of the difference, as $(P(x_0 + x_1), P(x_1 + x_2), P(x_2 + x_3) \ldots, P(x_{n-1} + x_0))$. We prove (following Caragiu et al. 2014) that the GPF Ducci iteration always enters a limit cycle, and if that cycle is nontrivial (of length greater than 1), then not only does every vector in the limit cycle have components in the set $\{2,3,5,7\}$, but every element in the set $\{2,3,5,7\}$ occurs as a component for some of the vectors in the limit cycle. The proof showcases, again, amazing properties of the 4-element magma $\{2,3,5,7\}$, which may be of interest to combinatorial number theorists. We also conducted a comprehensive computational analysis of the limit cycle statistics for the GPF Ducci games starting with a seed that can be assumed (without loss of generality) to belong to the set $\{2,3,5,7\}^N$ for $N \leq 8$. Note that The GPF Ducci iteration may be seen as a "nonassociative" quaternary one-dimensional cellular automaton. In the same section, we also discuss an interesting analogue of the Proth–Gilbreath conjecture, proved in 2013 (Caragiu et al. 2013). This is similar to the GPF Ducci recurrence, with the iteration acting on sequences.

In Section 3.7 we investigate interesting aspects related to the generation of nonassociative algebraic structures $\left(\Pi, f_{a,b}\right)$, where $\Pi$ is the set of primes, and

$$f_{a,b}(x, y) = P(ax + by).$$

Then $\left(\Pi, f_{a,b}\right)$ is a magma structure (generally nonassociative and noncommutative). An interesting question raised by the author and Paul A. Vicol (graduate student at Simon Fraser University) was the possibility of the cyclicity of $\left(\Pi, f_{a,b}\right)$. For example, if $a = 2, b = 1$, then the structure $(\Pi, *) := \left(\Pi, f_{2,1}\right)$ is conjectured to be cyclic. If this conjecture is true, the every prime can be written using 2, the operation symbol, and parentheses. For example:

$$3 = 2 * 2$$
$$5 = 2 * (2 * (2 * (2 * 2)))$$
$$7 = 2 * (2 * 2)$$
$$11 = 2 * (2 * (2 * 2))$$
$$13 = (2 * 2) * (2 * (2 * 2))$$

$$\cdots\cdots\cdots\cdots\cdots$$

Computational evidence for the cyclicity conjecture was obtained with a Julia program running on Google Compute Engine servers (see Appendix C). A sequence of necessary conditions for the cyclicity of magma structures of the form $\left(\Pi, f_{a,b}\right)$ is presented.

Relatively recently, we decided to adapt our work by shifting from the greatest prime factor function to Conway's subprime function:

$$C(x) = \begin{cases} x, & \text{if } x \text{ is prime or } x = 1, \\ x/L(x), & \text{else} \end{cases}$$

where $L(x)$ is the least prime factor of $x$.

Conway's subprime function and related sequences and structures are discussed in Chapter 4. This function has already produced interesting contributions, among them the excellent 2014 *Mathematics Magazine* analysis of Conway's subprime Fibonacci sequences by R.K. Guy, T. Khovanova, and J. Salazar. In a recent joint work with Paul A. Vicol, we managed to prove that if we define the binary operation $x \oplus y := C(x+y)$ on the set of natural numbers $\mathbb{N} = \{1, 2, 3, \ldots\}$, then the magma structure thus obtained is indeed cyclic:

$$\mathbb{N} = \langle 1 \rangle.$$

Moreover, if the sets $\{C_n\}_{n \geq 0}$ are defined, as usual, by $C_0 = \{1\}, C_{n+1} = C_n \cup (C_n \oplus C_n)$, then the cardinalities $|A_n|$ have been computed up to $n = 32$ (with the last step taking more than a week of running Julia on a Google server), and computational evidence has been provided in support of the amazing conjecture

$$\lim_{n \to \infty} \frac{|C_{n+1}|}{|C_n|} = \frac{1 + \sqrt{5}}{2}.$$

Moving on to sequences, computational evidence indicates that the ultimate periodicity conjectured for Conway's subprime Fibonacci sequences satisfying $x_n = C(x_{n-1} + x_{n-2})$ may be extended to higher-order analogues, i.e., sequences of natural numbers satisfying $x_n = C(c_1 x_{n-1} + c_2 x_{n-2} + \cdots + c_d x_{n-d} + c_0)$. For example, in the Conway tribonacci class of sequences (see Section 4.3) satisfying $x_n = C(x_{n-1} + x_{n-2} + x_{n-3})$, we found periods 3174, 6, and 5. The following graph displays a Conway tribonacci sequence beginning with 0, 0, 1. The length of the limit cycle will be 3174, while the maximum element in the cycle is 454,507.

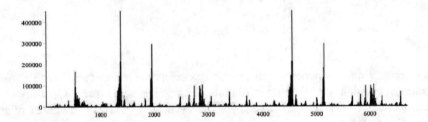

In Section 4.4 we study Conway subprime Ducci games, which are similar to the GPF Ducci games, but with the recurrence given by

$$(x_1, x_2, \ldots, x_n) \to (C(x_0 + x_1), C(x_1 + x_2), \ldots, C(x_{n-1} + x_0)).$$

A Floyd–Monte Carlo analysis of the limit cycles of the Conway Ducci recurrence in dimensions up to 8 is provided in Section 4.4. The "Floyd–Monte Carlo method" is one that we used fairly often for the exploration of the period space and the relative frequencies of the occurring periods (see Section 2.4 and Appendix B) for various integer or vector sequences based on the greatest prime factor function, or on Conway's subprime function.

Several possible applications of the recurrent sequences and structures considered here, based on either the greatest prime factor function or Conway's subprime function, are discussed in the largely experimental Chapter 5.

In the first two sections of the chapter we introduce and explore classes of nonassociative 2D cellular automata: some of them, based on the greatest prime factor section, may be viewed as 2D analogues of the (one-dimensional) GPF Ducci games and are quaternary, with the possible states of a cell being 2, 3, 5, and 7. Others have more ample sets of states (albeit finite if we assume the ultimate periodicity conjecture for multidimensional GPF sequences) and feature an amazing complexity in their evolution.

The last two sections of Chapter 5 are part of an interesting ongoing project. In Section 5.3, we build a class of $\pm 1$ walks based on GPF sequences. The original intent was to show that they are good models of genuine random walks. On a short scale (if we look at a fixed number of initial terms for GPF sequences with sufficiently large recurrence coefficients, sufficiently large recurrence degree, and sufficiently large components of the seed), they appear to behave like random walks from the point of view of the displacement statistics and a grouping test. But on a larger scale, computational data begged to differ (even if they pass a certain quasirandomness test) and we consider ourselves lucky to have noticed an interesting phenomenon: the displacements of the $\pm 1$ walks that we introduce are negative, something that comes strangely close to the so-called Chebyshev bias (to the effect that even if the primes congruent to 3 modulo 4 are asymptotically equal in number to the primes congruent to 1 modulo 4, it appears that in the "race" between the two congruence classes of primes, those congruent to 3 modulo 4 are more often on top). While we cannot link directly the negative displacement in GPF walks with Chebyshev bias, a very exciting theoretical problem is thus raised.

In Section 5.4 we introduce a class of bitstreams (0/1 sequences) based on GPF sequences (by assigning a "0" to the terms of the GPF sequence that are either 2 or congruent to 1 modulo 4, and a "1" to the terms that are congruent to 3 modulo 4) and analyze them from the point of view of linear complexity. We pay special attention to the infinite bitstreams obtained by repeating the bits associated with the limit cycle of a GPF sequence. The good news is their surprisingly good linear complexity, and generally large periods (even if the order of the underlying hidden GPF recurrence is relatively small), which makes them potentially useful as stream ciphers for one-time pad cryptography. That would mean that a problem of "$3x + 1$" type (which some consider to be simply mathematical curiosities, even if difficult to solve) has found an important application (especially given today's context).

# Chapter 2
# Warming Up: Integers, Sequences, and Experimental Mathematics

Motivational: quick overview, exciting trivia, second looks,
startups ....

## 2.1 From the Lebombo Bone to OEIS

The oldest known record of a mathematical object is the fossilized Lebombo bone [more than 43,000 years old according to rigorous carbon dating (d'Errico et al. 1987)], displaying 29 tally marks. Just for fun: this means that the first integer ever recorded in human history happens to be a prime number! One could indeed say that the natural numbers 1, 2, 3, ... have fascinated humanity from the dawn of time, thus becoming engrained in human consciousness (Greathouse; Dehaene 1999).

Integer sequences are mathematical objects with a long history. Babylonian mathematics is a culture displaying interesting computationally oriented features, some involving sequences, as exemplified by the sequence of squares of numbers up to 59 and the sequence of cubes of numbers up to 32 appearing in clay tablets from Senkerah dating from 2000 B.C. (Allen 2016).

An interesting integer sequence is a pattern waiting to be discovered (and hence an interesting problem to be solved) and an evolving phenomenon to be witnessed, which appears to have a life of its own. It elicits passion and interest from the inquisitive, creative mind, opening it to an experiential approach in mathematics. And for undergraduate students, such sequences, in the author's view, constitute one of the most powerful tools for advancing a high-impact learning in many mathematical areas, especially in number theory and its applications.

Arguably the most important resource on integer sequences today, luckily available for free, is the On-Line Encyclopedia of Integer Sequences (OEIS), http:// oeis.org. Initiated by Neil J.A. Sloane in early 1964 (OEIS; NJA Sloane 2013) and continuously maintained and expanded by a group of enthusiasts and contributors, OEIS is a huge list of interesting sequences. Every one of them is identified by an OEIS code. For example, just to mention four classical examples, the sequence of

© Springer International Publishing AG 2017
M. Caragiu, *Sequential Experiments with Primes*,
DOI 10.1007/978-3-319-56762-4_2

natural numbers (1, 2, 3, 4, 5, …) is listed as A000027, the sequence of primes (2, 3, 5, 7, 11, …) goes under the label A000040, while the sequence of Fibonacci numbers (0, 1, 1, 2, 3, 5, 8, …, i.e., every term is the sum of the previous two, starting from the "seed" 0, 1) goes under the OEIS code A000045. Every listing is accompanied by a rich and useful set of related comments, references to published work, notes, links, theorems, available explicit formulas, program lines (MAPLE, Mathematica, etc.) generating the sequence, and a list of OEIS codes of closely related sequences. OEIS is a valuable research resource for the mathematical community including undergraduate students aspiring to perform meaningful work in, or with, mathematics.

## 2.2  Experimental Mathematics

OEIS lists, with detailed references, an amazing number of new sequences at the forefront of current research. N.J.A. Sloane is right in calling it "an endless source of open problems." This definitely makes OEIS an important player and a useful collaborative tool in the emerging area known as experimental mathematics.

On the University of Copenhagen web page dedicated to "Experimental Mathematics in Number Theory, Operator Algebras, and Topology" (ExpMathDK 2016) can be found the following concise definition:

> Experimental mathematics is the modus operandi of using computers for generating insight into pure mathematics.

Kurt Gödel, the mathematical visionary, anticipated—in an unpublished 1951 essay—a deep connection between the approaches taken in physics and experimental mathematics, with an interesting philosophical overtone [in the platonic view, mathematical truth is objectively "out there" to be discovered by the working mathematician (Gödel 1951)]:

> If mathematics describes an objective world just like physics, there is no reason why inductive methods should not be applied in mathematics just the same as in physics.

We can say that from an experimental mathematics viewpoint, we learn to experiment with numbers (say integer sequences) and see how they behave. The hope is that interesting patterns and phenomena will appear along the way. Especially when more computing power is involved, thus making possible the manipulation of larger integers, it is easier to accept (with an eye to the above-mentioned Gödelian analogy with physics) that new emergent "mathematical phenomena" will pop up, pretty much like when new elementary particles and related phenomena arguably appear as a result of higher-energy collisions in more powerful particle accelerators.

Of course, computer experimentation is rewarding in itself as a potentially limitless exploration of the (mathematical) space. Far from being a blind walk, experimentation/computation in mathematics is rather a process of meaningful intentionality, by creatively engaging the human person, with the hope that amazing facts will present themselves. This is, in fact, the destiny of the working mathematician, so eloquently and artfully presented by Frenkel (2013):

> Engaging implies hard work, More often than not, at the end of the day (or a month, or a year), you realize that your initial idea was wrong, and you have to try something else. These are the moments of frustration and despair. You feel that you have wasted an enormous amount of time, with nothing to show for it. This is hard to stomach. But you can never give up. You go back to the drawing board, you analyze more data, you learn from your previous mistakes, you try to come up with a better idea. And every once in a while, suddenly, your idea starts to work. It's as if you had spent a fruitless day surfing, when you finally catch a wave: you try to hold on to it and ride it for as long as possible. At moments like this, you have to free your imagination and let the wave take you as far as it can. Even if the idea sounds totally crazy at first.

The implicit expectation is that the patterns and structures that appear in this experimental/computational process of discovery

- will receive proofs (or refutations for that matter) somewhere down the line (one may try partial proofs first), and
- will inspire in real time further meaningful experiments, adding to the wonder and the complexities of the problem.

Last but not least, if an interesting "artistic" pattern emerges in the process, it would be helpful to dwell on it further, since there is a good reason, waiting to be discovered, behind the observed symmetry. Also, you may want to hold on to such discoveries: they are good advertising for mathematics.

## 2.3  Primes

With an experimental mathematics mindset, lets us turn back to A000040 (the sequence of primes, i.e., positive integers greater than 1 without any divisor other than 1 and themselves):

$$2, 3, 5, 7, 11, 13, 17, 19, 23, 29, 31, 37, 41, 43, 47, 53, 59, 61, 67, 71, 73, 79, 83, 89, 97, \ldots.$$

This is a tremendously important sequence for mathematics as a whole. Without an immediately discernible pattern, they form a rich data set that has fascinated mathematicians for millennia. In reference to the sequence of primes, Tim Gowers said (2002):

> Although the prime numbers are rigidly determined, they somehow feel like experimental data.

Crucial results on primes go back to the Greek mathematician Euclid's *Elements* (2016) (about 300 BC). Among them:

- There are infinitely many primes.
- Every positive integer has a unique prime factorization (which makes prime numbers the multiplicative building blocks for the set of natural numbers).
- The importance of primes of a special form: thus, the number $2^{n-1}(2^n - 1)$ is perfect (equals the sum of its proper divisors) whenever $2^n - 1$ is prime.

Fast-forwarding to today, arguably the single most important mathematical object that monolithically encodes the sequence of primes (and is the decisive factor in the study of the distribution of primes) is Riemann's zeta function (Edwards 1974):

$$\varsigma(s) = \sum_{n=1}^{\infty} \frac{1}{n^s} = \prod_{p \, \text{prime}} \frac{1}{1 - \frac{1}{p^s}}.$$

The zeta function can be analytically continued to a complex function that is holomorphic at every point except $s = 1$ (where it has a simple pole). The seminal 1896 result known as the prime number theorem, proved independently by Jacques Hadamard and Charles Jean de la Vallée-Poussin through extending Riemann's ideas, gives an asymptotic form of the "prime counting function" $\pi(x) = \#\{p | p \text{ prime}, p \leq x\}$:

$$\lim_{x \to \infty} \frac{\pi(x)}{x / \ln x} = 1.$$

Some consequences of $\pi(x) \approx x / \ln x$:

- The asymptotic formula for the $n$th prime: $p_n \approx n \ln n$. This can be refined by the inequality $n(\ln n + \ln \ln n - 1) < p_n < n(\ln n + \ln \ln n)$, holding for $n \geq 6$ (Rosser and Schoenfeld 1962; Dusart 1999; Bach and Shallit 1996).
- For every given positive $\varepsilon$, the interval $(n, n + \varepsilon n)$ contains a prime, provided $n$ is large enough.

Sometimes, for computational purposes, especially when primes in small intervals are involved, we need *effective bounds*. A classical one is Bertrand's postulate, proved in 1850 by Chebyshev:

*Chebyshev's theorem:* For every $n > 1$ there exists a prime $p$ such that $n < p < 2n$.

There are many more powerful results in the same "effective" category. We will refer to only two of them, just to give the reader a "taste":

- For $x \geq 25$ there is a prime $p$ with $x < p < 1.2x$ (Nagura 1952).
- For $x \geq 3275$ there is a prime $p$ with $x < p < x\left(1 + \frac{1}{2(\ln x)^2}\right)$ (Dusart 2016).

Under a plausible but still unproven assumption (the Riemann hypothesis), the function $\pi(x)$ allows for the sharper estimate $\pi(x) = li(x) + O(\sqrt{x} \ln x)$ (see Davenport 2000, p. 113), where $li(x) = \int_2^x \frac{dt}{\ln t}$ can be estimated (see Davenport 2000, p. 54), for fixed $q$ and $x \to \infty$, as $li(x) = \frac{x}{\ln x} + \frac{1!x}{(\ln x)^2} + \cdots + \frac{q!x}{(\ln x)^{q+1}}[1 + \varepsilon(x)]$.

### 2.3.1 Harmonic Numbers Revisited

For $n \geq 2$, the partial sums $H_n := \sum_{k=1}^{n} \frac{1}{k}$ (harmonic numbers) of the harmonic series are never integers. Having just discussed Chebyshev's theorem, it is worth noticing a nice immediate consequence. For $n \geq 2$, let $p$ be the largest prime $p \leq n$. Then $2p > n$ (otherwise, if $2p \leq n$, Chebyshev's theorem implies the existence of a prime $q$ with $p < q < 2p \leq n$, contradicting the choice of $p$). Thus $\frac{1}{p}$ is the only summand in $H_n$ that is divisible by $p$. That excludes the possibility of $H_n$ being an integer, since in fraction form, $H_n$ would have a denominator divisible by $p$ (to exactly the first power) with a numerator not divisible by $p$.

Of course, the conclusion $H_n \notin \mathbb{Z}$ for $n \geq 2$ is traditionally reached without Chebyshev's theorem, by considering the (necessarily unique) summand $\frac{1}{k}$ that is divisible by the highest power of 2.

But the use of Chebyshev's theorem also implies that $H_n$ cannot be a power of order greater than 1 (square, cube, etc.) of a rational number, and also that if $\pi(m) < \pi(n)$, then $H_n - H_m$ cannot be a power of order greater than 1 of a rational number.

Note that the above-mentioned refinement of Chebyshev's theorem due to Nagura (1952) can be used to prove the following proposition (Bax and Car 2004).

**Proposition 2.1** *For $n \geq 2$, the sum $\sum_{k=1}^{n} \frac{1}{k^k}$ cannot be a perfect $r$-th power $(r \geq 2)$ of a rational number.*

*Proof* Note that Nagura's result about the existence of a prime between $x$ and $\frac{6}{5}x$ for $x \geq 25$ consequently implies that for $n \geq 11$, there are at least two distinct primes $p, q$ satisfying $\frac{n}{2} < p < q < n$. Then it is easy to see that in its reduced form, $\sum_{k=1}^{n} \frac{1}{k^k}$ is necessarily of the form $S_n := \sum_{k=1}^{n} \frac{1}{k^k} = \frac{A}{p^p q^q B}$, with $p, q$ not appearing in $AB$. If we assume $S_n$ to be an $r$th power $(r \geq 2)$, then the exponents of all primes appearing in the reduced fraction form of $S_n$ must be divisible by $r$, so that both $p$ and $q$ are multiples of $r$, a contradiction. This finishes the proof for $n \geq 11$. For $n \leq 10$, we will proceed with "brute force" computation (actually not so "brute," given the strong computational capabilities of MAPLE):

```
> for n from 2 by 1 to 10 do S(n):=sum(1/k^k,k=1..n): end do:
seq(ifactor(S(n)),n=2..10);
>
```

$$\frac{(5)}{(2)^2},\ \frac{(139)}{(2)^2\,(3)^3},\ \frac{(8923)}{(2)^8\,(3)^3},\ \frac{(27891287)}{(2)^8\,(3)^3\,(5)^5},\ \frac{(29)\,(431)\,(60251)}{(2)^8\,(3)^6\,(5)^5},$$

$$\frac{(49739)\,(2153)\,(2617)\,(2213)}{(2)^8\,(3)^6\,(5)^5\,(7)^7},\ \frac{(1173145819)\,(305147)\,(113539)}{(2)^{24}\,(3)^6\,(5)^5\,(7)^7},$$

$$\frac{(31)\,(1161244100781371)\,(600034207)}{(2)^{24}\,(3)^{18}\,(5)^5\,(7)^7},$$

$$\frac{(43)\,(37994429709572209)\,(16926029)\,(2441)}{(2)^{24}\,(3)^{18}\,(5)^{10}\,(7)^7}$$

As one can easily see from the above factorizations, $S_n\ (2 \le n \le 10)$ is not a perfect power of a rational number of order greater than 1.

Computer algebra platforms such as MAPLE and MATLAB are widely available in colleges today. If one needs a large prime, the MAPLE function **nextprime (x)** will provide the smallest prime greater than $x$ [after one has loaded MAPLE's number theory package using **with(numtheory)**]:

```
> with(numtheory):
nextprime(651456484752135412837652138251542874585 42);
                 6514564847521354128376521382515428 74585651
```

Integer factorization is also straightforward in MAPLE (with **ifactor**),

```
> ifactor(872659817561785987561985 61);
            (3) (47) (127) (48732887561388 61828123)
```

as is the determination whether a given positive integer is prime (with **isprime**),

```
> isprime(8721659812765129876519);
```
$$true$$
$$false$$

the $n$th prime (**ithprime(n)**),

```
> ithprime(100);
```
$$541$$

and the sequence of the first $n$ primes, for example for $n = 100$,

```
> seq(ithprime(k),k=1..100);
```
   2, 3, 5, 7, 11, 13, 17, 19, 23, 29, 31, 37, 41, 43, 47, 53, 59, 61, 67, 71, 73, 79, 83, 89, 97, 101,
       103, 107, 109, 113, 127, 131, 137, 139, 149, 151, 157, 163, 167, 173, 179, 181, 191, 193,
       197, 199, 211, 223, 227, 229, 233, 239, 241, 251, 257, 263, 269, 271, 277, 281, 283, 293,
       307, 311, 313, 317, 331, 337, 347, 349, 353, 359, 367, 373, 379, 383, 389, 397, 401, 409,
       419, 421, 431, 433, 439, 443, 449, 457, 461, 463, 467, 479, 487, 491, 499, 503, 509, 521,
       523, 541

Frequently used in introductory courses in number theory or discrete mathematics are two algorithms of fundamental importance in cryptography applications:
   ***The extended Euclidean algorithm (EEA)*** expresses the greatest common divisor $\gcd(a,b)$ as a linear combination $ax+by$ with $x,y \in \mathbb{Z}$. For example, $\gcd(13,11) = 1 = 13 \cdot (-5) + 11 \cdot 6$. In MAPLE we may proceed like this (try it):

```
> igcdex(13,11,'x','y'); x; y;
```

Students may want first to get used to performing the EEA with paper and pencil, by building a table with four columns, labeled $r$ (remainder), $q$ (quotient), $x$, and $y$. The first two rows (besides the head) are just "data entry," with the numbers $a, b$ with $a > b$ entered under the remainder column in decreasing order. Say $a = 139, b = 49$. The column labeled $x$ is associated with the larger number, while the column labeled $y$ is associated with the smaller number. At every stage, the number in the remainder column must be equal to $a$ times the number in the $x$ column plus $b$ times the number in the $y$ column. Afterward, updates will be performed using the previous two rows, as follows:

- Divide the bigger remainder by the smaller one; enter the quotient in the $q$ column.
- Multiply the newly obtained $q$ by the prior $(x, y)$ vector, subtract this from the $(x, y)$ vector obtained two steps before, and enter the result as the updated $(x, y)$ vector.
- The numbers in the remainder column will form a decreasing sequence. Continue to update until you see zero in the remainder column. The last nonzero remainder is $d = \gcd(a, b)$, which will be found in addition to the desired $x$, $y$ such that $d = ax + by$.

The EEA table below leads to $\gcd(139, 49) = 1 = 139 \cdot 6 + 49 \cdot (-17)$:

| $r$ | $Q$ | $x$ | $y$ |
| --- | --- | --- | --- |
| 139 | – | 1 | 0 |
| 49 | – | 0 | 1 |
| 41 | 2 | 1 | –2 |
| 8 | 1 | –1 | 3 |
| 1 | 5 | 6 | –17 |
| 0 | … | … | … |

*Fast modular exponentiation* calculates modular powers of large integers using the method of repeated squaring (symbolized by "&^", as opposed to "^", which would first attempt to calculate the exact integer value of the exponential (resulting in an obvious overflow error):

> 76456435614965&^625623456435435 mod

9324859243852943872493368794238679234;

  1456351055027143913768634745274150553

Again, before we can experience the sophisticated computing power of MAPLE, it would be quite rewarding for us (both faculty and students) to have, again, a paper and pencil experience of this. A hands-on method for the fast calculation of $a^N$ mod $m$ involves a table with two columns. At the head of the table we have, in the left column, $k$, the "repeated squaring exponent," which takes power of 2 values $1, 2, 4, 8, 16, \ldots$ and continues downward for as long as $k \leq N$. At the head of the right column we will put $a^k$ mod $m$. We initialize the first row with $1(= k)$ in the first column, and $a(= a^1)$ in the second column. To update:

- In the first column, double the exponent (previous $k$ value).
- In the second column, square the previous entry modulo $m$.
- Continue for as long as the item in the left column is $\leq N$.

Say, for example, that $a = 23$, $N = 100$, $m = 541$. The table shapes up as follows:

| $k$ | $a^k$ mod $m$ |
|-----|---------------|
| 1   | 23            |
| 2   | 529           |
| 4   | 144           |
| 8   | 178           |
| 16  | 306           |
| 32  | 43            |
| 64  | 226           |

The first stage is complete. Now we have to write the exponent 100 as a sum of entries in the left column. This can be done using a base 2 representation of 100, to the effect that $100 = 64 + 32 + 4$. Thus $23^{100} = 23^{64} \cdot 23^{32} \cdot 23^4$ mod $541 = 226 \cdot 43 \cdot 144 = 366$. Not comfortable with bases of numeration? That's OK. We can substitute this by a "greedy" technique of writing the exponent as a sum of powers of 2, by repeatedly removing the highest power of 2 from what's left. Say we use this for 100. The highest power of 2 not greater than 100 is 64. Put it aside and continue with the leftover, 36. The highest power of two not greater than 36 is 32. Put it aside, and we are left with a 4, which will be put aside in the next step, so we have $100 = 64 + 32 + 4$. It is perhaps tedious, but also rewarding and beautiful (albeit slow), but in this way we will probably appreciate even more the depth involved in such specialized computer algebra software and be able to enjoy it fully.

## 2.4 Periodic Sequences: Visualization, Periods, Preperiods, Floyd's Cycle-Finding Algorithm

The terms of the sequences investigated in this book will be defined in terms of prime numbers. When it comes to sequences, especially when one is looking for periodicity, it is useful to have a look at a plot of the sequence so that we can immediately become aware of any obvious repetitive patterns.

```
> x(0):=0; x(1):=1; for k from 2 to 100 by 1 do x(k):=(x(k-
1)+x(k-2)) mod 29 end do: L:=[seq(x(k),k=0..100)]: listplot(L);
```

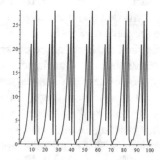

This represents a plot of the Fibonacci numbers modulo 29. The period is 14, and this can be discovered "low-tech" by gradually inspecting the picture and possibly checking neighborhoods of a few terms around subsequent repeating peaks and valleys.

This looks like an exciting adventure, but when it comes to more complex periodic sequences of integers (or integer vectors), a more formal and precise algorithm for determining the period is needed. This is Floyd's algorithm for cycle detection, also known as *the tortoise and the hare algorithm*; cf. Knuth (1969, p. 7). This cycle-detection method is instrumental in Pollard's rho randomized factoring algorithm (Pollard 1975).

For an ultimately periodic sequence $x_0, x_1, x_2, x_3, x_4, \ldots$ satisfying a recurrence $x_{i+1} = f(x_i)$ with period $\lambda$ and preperiod $\mu$ (so that $x_i = x_{i+\lambda}$ for all $i \geq \mu$, but not earlier), we proceed with two sequence calculations: *a calculation at* "slow" speed (also known as the *tortoise*) $t_i = x_i$, $i = 0, 1, 2, \ldots$, will be synchronously produced and compared with the terms of a "fast" calculation (the *hare*) $h_i = x_{2i}$, $i = 0, 1, 2, \ldots$. That is, at every step, we will compare $x_i$ with $x_{2i}$. Note that the distance between the subscripts, $i$, that is, the distance between the hare at the front and the tortoise in the back, increases by 1 at every step.

The point of the algorithm is the following: *once both tortoise and hare enter the periodicity regime* $(i \geq \mu)$, due to the gradual increments, in steps of 1, of the distance between them, the distance will at some point become an integer multiple of $\lambda$, in which case $x_i = t_i = h_i = x_{2i}$. It is important to notice that the form of the recurrence (iteration of a certain function) has as a consequence that no term of the sequence that occurs prior to the start of the cycle is equal to any one of the terms in the cycle, so that $i \geq \mu$ implies $x_m \neq x_n$.

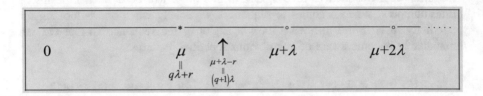

Let $q$ and $r$ be integers such that $\mu = q\lambda + r, 0 \leq r < \lambda$ (keep in mind that we don't assume any a priori information about $\lambda$ and $\mu$). It is easy to see that the first index $i$ right after $\mu$ with the property $x_i = x_{2i}$ is $i = \mu + \lambda - r$ (that is, $(q+1)\lambda$). At this moment when the tortoise and the hare can be construed as projecting onto the same point of the circle of circumference $\lambda$, we have a moment of "collision." Due to the gradual increment of the hare–tortoise distance, the next value of $i > \mu + \lambda - r$ for which $x_i = x_{2i}$ will occur after $\lambda$ more steps, for $i = \mu + 2\lambda - r = (q+2)\lambda$, that is, $x_{\mu + 2\lambda - r} = x_{2\mu + 4\lambda - 2r}$. Indeed, while the tortoise goes once around the circle, the hare does it twice, so they meet again after an increment of $\lambda$ in the subscript: the difference between the tortoise's (that is, $x_i$) subscripts $i$ at the first two points of collision will be $(\mu + 2\lambda - r) - (\mu + \lambda - r) = \lambda$. Thus we obtain the period $\lambda$.

For the preperiod $\mu$, starting from the instance of the first collision ($x_i = x_{2i}$ with $i = \mu + \lambda - r$) we will go back at most $\lambda$ times ($\lambda$ is known at this time), more exactly $\lambda - r$ times, and keep testing $x_i = x_{i+q}$, $x_{i-1} = x_{i-1+q}$, $x_{i-2} = x_{i-2+q}, \ldots$ with the last instance of continuing equality ($x_\mu = x_{\mu+q}$) providing the desired preperiod $\mu$. Let us agree to call this particular $i = \mu + \lambda - r$ the "tentative start" for the series of equality checks going back at most $\lambda$ times.

Note an important byproduct of the above discussion: the "tentative start" is a multiple of the period.

Let us apply the ideas in the above discussion to concrete problems in which MAPLE will prove very useful. The first example involves a nonlinear first-order recurrent sequence in modular arithmetic (the type of sequences used in Pollard's factoring method):

*Example 2.1* Consider the sequence of integers $\{x_n\}_{n \geq 0}$ defined as follows:

$$(*)\begin{cases} x_0 = 12345, \\ x_{n+1} = x_n^2 + 1 \bmod 4429. \end{cases}$$

We will write down the series of MAPLE commands that lead us to the period/preperiod calculation. First of all, we will produce the "tentative start" ($i = \mu + \lambda - r$ in the discussion above, that is, the first instance in which both tortoise and hare are in the periodicity regime, while the distance between them will be an integer multiple of the period):

```
> with(numtheory): with(ListTools): N:=12345; f:=x->x^2+3 mod N;

> x:=25: SLOW:=f(x): FAST:=f(f(x)): k:=1: while SLOW<>FAST do
SLOW:=f(SLOW): FAST:=f(f(FAST)): k:=k+1; end do:
TENTATIVE_START:=k;
```
$$TENTATIVE\_START := 21$$

Second, we will calculate the period by considering when tortoise and hare (now both in the cycle) meet again:

```
> x:=SLOW: SLOW:=f(x): FAST:=f(f(x)): k:=1: while SLOW<>FAST do
SLOW:=f(SLOW): FAST:=f(f(FAST)): k:=k+1; end do: PERIOD:=k;
```
$$PERIOD := 7$$

This means that the period is 7 and that the periodic regime starts at most 7 steps before 21.

For the verification step, the following relevant terms of the sequence will be produced:

```
> t(0):=25: for r from 1 to TENTATIVE_START+PERIOD do
t(r):=f(t(r-1)): end do:
X:=seq(t(r),r=0..TENTATIVE_START+PERIOD);
```
$X := 25, 628, 11692, 6682, 9607, 3232, 1957, 2902, 2317, 10762, 12202, 8107, 11017, 10597,$
$6292, 11197, 9337, 11527, 2497, 787, 2122, 9307, 7732, 9337, 11527, 2497, 787, 2122,$
$9307$

Thus the preperiod is $\mu = 16$, which translates into the following periodicity property of the sequence defined by $(*) : x_{n+7} = x_n$ for every $n \geq 16$ (and not earlier). As we shall see below, the trio "tentative start + period + preperiod" can be incorporated into a single block of MAPLE code (although for the exploration of the space of the possible periods of a certain recurrence relation, only the "period" part will be used). Before that, yet another example.

The second example will deal with second-order recurrences, more specifically the sequence of Fibonacci numbers modulo 55.

*Example 2.2* Consider the sequence of integers $\{x_n\}_{n \geq 0}$ defined as follows:

$$(**) \quad \begin{cases} x_0 = 0, x_1 = 1, \\ x_n = x_{n-1} + x_{n-2} \, 55 (n \geq 2). \end{cases}$$

In order to adapt the previous calculations to second-order recurrences, we will treat a second-order recurrence as a first-order recurrence, albeit with 2D vectors:

$$(x_n, x_{n+1}) \rightarrow (x_{n+1}, x_n + x_{n+1} \bmod 55)$$

```
> with(numtheory): with(ListTools): f:=(x,y)->(y,x+y mod 55);
```
$$f := (x,y) \rightarrow (y, (x+y) \bmod 55)$$

```
> X:=(0,1): SLOW:=f(X): FAST:=f(f(X)): k:=1: while SLOW<>FAST do
SLOW:=f(SLOW): FAST:=f(f(FAST)): k:=k+1; end do:
TENTATIVE_START:=k;
```
$$TENTATIVE\_START := 20$$

```
> X:=SLOW: SLOW:=f(X): FAST:=f(f(X)): k:=1: while SLOW<>FAST do
SLOW:=f(SLOW): FAST:=f(f(FAST)): k:=k+1; end do: PERIOD:=k;
```
$$PERIOD := 20$$

```
> t(0):=0: t(1):=1: for r from 2 to TENTATIVE_START+PERIOD do
t(r):=t(r-1)+t(r-2) mod 55 end do:
X:=[seq(t(r),r=0..TENTATIVE_START+PERIOD)];
```
$$X := [0, 1, 1, 2, 3, 5, 8, 13, 21, 34, 0, 34, 34, 13, 47, 5, 52, 2, 54, 1, 0, 1, 1, 2, 3, 5, 8, 13, 21, 34, 0,$$
$$34, 34, 13, 47, 5, 52, 2, 54, 1, 0]$$

In this case, the conclusion is that the Fibonacci sequence modulo 55 has period 20 with a preperiod of 0.

Faced with the case of a particular recurrent sequence $x_n = f(x_{n-1}, \ldots, x_{n-d})$ of a certain order $d$, with the initial conditions constituting the "seed" vector $SEED := (x_0, x_1, \ldots, x_{d-1})$ already in place, and keeping in mind that the "tentative start" is an integer multiple of the period, as observed previously, we shall use the following compact block of MAPLE code to derive all three, in that order:

- Tentative start
- Period
- Preperiod

We have seen above how to produce the first two. The idea for the preperiod is to start with the vectors ($d$-tuples), namely the initial $d$-tuple $U := "SEED"$ and $V$ (the $d$-tuple found after a shift of "tentative start" ahead), setting a variable $m$ with an initial value of zero, proceeding with the shifting of the above-mentioned $d$-tuples $U$ and $V$, one step at a time, via the process

$$(x_k, x_{k+1}, \ldots, x_{k+d-1}) \mapsto (x_{k+1}, \ldots, x_{k+d-1}, f(x_{k+d-1}, x_{k+d-2}, \ldots, x_{k+1}, x_k)),$$

while increasing $m$ by 1 for each shift, *for as long as U and V remain distinct*. This will continue until we get to the situation $U = V$ (which must happen, and when it happens for the first time, $m = PERIOD$. Let's take, for example, the third-order recurrence

$$\begin{cases} x_n = P(x_{n-1} + x_{n-2} + x_{n-3}), \\ x_0 = 11, x_1 = 3, x_2 = 37. \end{cases}$$

where $P(x)$ is the greatest prime factor of $x$ (an important function for our purposes, which will become one of our heroes a little bit later; the essential thing needed here is that there are reasons to believe that a whole class of recurrence relations similar to this one produce ultimately periodic sequences).

```
> f:=(x1,x2,x3)->(x2,x3,P(x3+x2+x1)): SEED:=(11,3,37): X:=SEED:
SLOW:=f(X): FAST:=f(f(X)):   k:=1: while SLOW<>FAST do
SLOW:=f(SLOW): FAST:=f(f(FAST)): k:=k+1; end do:
TENTATIVE_START:=k; Y:=SLOW: X:=SLOW: SLOW:=f(X): FAST:=f(f(X)):
k:=1: while SLOW<>FAST do SLOW:=f(SLOW): FAST:=f(f(FAST)):
k:=k+1; end do: PERIOD:=k; m:=0: U:=SEED: V:=Y: while U<>V do
m:=m+1; U:=f(U); V:=f(V) end do: PREPERIOD:=m;
```

We get, fairly quickly,

$$
\begin{aligned}
\text{TENTATIVE\_START} &:= 400 \\
\text{PERIOD} &:= 100 \\
\text{PREPERIOD} &:= 346
\end{aligned}
$$

Again, let's say that we don't believe it until we see it (which is natural for amateur programmers like the author). Let's get enough terms:

```
> x(0):=11: x(1):=3: x(2):=37: for r from 3 to 500 do
x(r):=P(x(r-1)+x(r-2)+x(r-3)): end do: L:=[seq(x(r),r=0..500)]:
```

Then by directly selecting suitable blocks of three consecutive terms, we indeed obtain the confirmation of the Floyd algorithm's above "triple output":

$$(x_{346}, x_{347}, x_{348}) = (5, 13, 7) = (x_{346+100}, x_{347+100}, x_{348+100}),$$
$$(x_{345}, x_{346}, x_{347}) = (17, 5, 13) \neq (31, 5, 13) = (x_{345+100}, x_{346+100}, x_{347+100}).$$

In many cases, especially when we will investigate (through a sort of "Monte Carlo–Floyd" hybrid) the distribution of the periods with initial conditions randomly varied, we are interested in the period only. This means that after specifying the recurrence form and the seed, we will use only the first part of the code that leads to the period:

```
> X:=SEED:  SLOW:=f(X): FAST:=f(f(X)):  k:=1: while SLOW<>FAST
do SLOW:=f(SLOW): FAST:=f(f(FAST)): k:=k+1; end do:
TENTATIVE_START:=k: Y:=SLOW: X:=SLOW: SLOW:=f(X): FAST:=f(f(X)):
k:=1: while SLOW<>FAST do SLOW:=f(SLOW): FAST:=f(f(FAST)):
k:=k+1; end do: PERIOD:=k;
```

**Exercise**: assume that after the "tentative start" and the "period" are identified, somebody proposes the following MAPLE procedure (based rather on "backtracking"):

```
> PP:=proc(X::list, TS::integer, P::integer);
> k:=TS;
> while X[k]-X[k+P]=0 do
> k:=k-1;
> end do;
> PP:=k;
> end proc;
```

Assume that somebody computes and stores the list

```
> X:=[seq(x(r),r=0..TENTATIVE_START+PERIOD)]:
```

Does the MAPLE instruction "PP(X, TENTATIVE_START, PERIOD)" provide the preperiod? When does it fail?

## 2.5  Mathematical Beauty at the Addition/Multiplication Interface

The sequence of natural numbers $1, 2, 3, 4, 5, 6, \ldots$ ($a_n = n$, the OEIS sequence A000027) is archetypal. Ideas involving the natural numbers evolved and became increasingly sophisticated over millennia, beginning with ancient tally marks and proceeding through the emergence of the sexagesimal numeration system in Mesopotamia.

With addition, the set of natural numbers $\mathbb{N} = \{1, 2, 3, 4, 5, 6, \ldots\}$ is a semigroup $(\mathbb{N}, +)$ generated by a single element, indeed $\mathbb{N} = \langle 1 \rangle$. Here is a method to analyze the "speed" of additive generation that will be applied later in the context of other special binary operations. Starting from $A_0 = \{1\}$, we define an ascending sequence of sets $\{A_n\}_{n \geq 0}$ through the recurrence $A_{n+1} = A_n \cup \{x + y | x, y \in A_n\}$.

That is, at every step, the current set $A_n$ is enlarged with all the values of the binary operation with arguments in $A_n$. It is not difficult to see that at each step, the set size doubles, i.e., $A_n = \{1, 2, 3, \ldots, 2^n\}$. We clearly have $A_0 \subset A_1 \subset A_2 \subset \ldots$ and $\langle 1 \rangle = \bigcup_{n=0}^{\infty} A_n = \mathbb{N}$.

As we shall see, if we replace addition with another well-chosen binary operation "∘" on $\mathbb{N}$ and repeat the above procedure for the structure $(\mathbb{N}, \circ)$, with sets $\{A_n\}_{n \geq 0}$ defined accordingly, the study of the speed of generation (specifically the limit $\lim(|A_{n+1}|/|A_n|)$ will lead us to an amazing conjecture that will provide a description of the golden section $(1 + \sqrt{5})/2$ in purely arithmetic terms.

In themselves, addition and multiplication on $\mathbb{N}$ are "easy." The prime factorization structures of two natural numbers $a, b$, if known (granted, that's a big "if," since factoring integers is a computationally hard problem), make the prime factorization of their product $ab$ immediately available. Indeed, the prime factorization theorem reduces the multiplication $ab$ to additions of the corresponding exponents of the various prime powers appearing in $a, b$: if $a = p_1^{\alpha_1} p_2^{\alpha_2} \ldots p_k^{\alpha_k}$ and $b = p_1^{\beta_1} p_2^{\beta_2} \ldots p_k^{\beta_k}$ with $\alpha_i, \beta_i \geq 0$, then $ab = p_1^{\alpha_1 + \beta_1} p_2^{\alpha_2 + \beta_2} \ldots p_k^{\alpha_k + \beta_k}$. In "fancier" terms, that can be alternatively formulated by saying that due to the prime factorization theorem, the multiplicative semigroup of positive integers is isomorphic to the direct sum of a countable family of identical additive monoids of the form $(\mathbb{N}_0, +)$, where $\mathbb{N}_0 = \{0, 1, 2, 3, 4, 5, \ldots\}$.

The really interesting problems appear, though, at the interface between additive and multiplicative aspects, a zone that can be rightfully considered the *Mariana Trench* of number theory. Through the process of integer factorization itself, an interesting connection was revealed (Billingsley 1973) with Brownian motion.

We can start by thinking that there is no immediate formula for the prime factorization of $a + b$ in terms of those of $a$ and $b$. Going only a little deeper within that interface will quickly lead us to difficult problems that are still open. If we look at the prime numbers as the pinnacles of the *multiplicative* universe, what happens when we start to *add* primes: $2 + 2 = 4$, $3 + 3 = 6$, $3 + 5 = 8$, $3 + 7 = 10$, $5 + 7 = 12$, and so on? It doesn't take long to notice the strong likelihood that every even number that is greater than 2 is a sum of two primes. This is the celebrated Goldbach conjecture, still unproven since it was stated on June 7, 1742, when Christian Goldbach formulated the conjecture in a letter to Leonhard Euler (Goldbach 2016).

Note that although it is important to be aware of various kinds of asymptotic behavior involving "primes at the additive–multiplicative interface," the proofs for most of them involve advanced analytic number theory and are beyond the scope of this book, which is limited to elementary approaches. That said, among many interesting problems emerging at this interface, we will mention the following.

The Sophie Germain primes, that is, primes $p$ such that $2p + 1$ is also a prime (OEIS sequence A005384, 2, 3, 5, 11, 23, 29, 41, 53, 83, 89, 113, 131, 173...). A heuristic result (Shoup 2009, p. 123) estimates the number of Sophie Germain primes less than or equal to $x$ to be a constant multiple of $x/(\ln x)^2$. A similar estimate exists for the number of primes $p$ less than or equal to $x$ such that $p + 2$ is also prime, i.e., $(p, p+2)$ form a twin prime pair. It has been proved (Brun's theorem, 1919) that the sum of the reciprocals of the primes in the twin pairs is convergent. The conjectural infinitude of both Sophie Germain primes and twin prime pairs is still unproven.

Primes in arithmetic progressions constitute another amazing phenomenon emerging at the additive–multiplicative interface. One of the most important number-theoretic results proved recently is the celebrated Green–Tao theorem (Green and Tao 2008), to the effect that there are arbitrarily long arithmetic progressions consisting of primes. The longest known such progression (discovered in 2010), due to Benoît Perichon, consists of 26 primes (see OEIS sequence A204189). As an example on the asymptotics side, in a seminal 2010 paper (Green and Tao 2010), Green and Tao estimated the number of four-term arithmetic prime progressions $1 < p_1 < p_2 < p_3 < p_4 \leq N$ as $(1 + o(1)) \cdot \frac{3}{4} \prod_{p \geq 5} \left(1 - \frac{3p-1}{(p-1)^3}\right) \frac{N^2}{\log^4 N}$, or approximately $\frac{0.4764 \cdot N^2}{\log^4 N}(1 + o(1))$.

On a related note, it is important to notice here the classical result (Dirichlet's theorem) on the existence of infinitely many primes in an arithmetic progression $\{qn + a | n = 0, 1, 2, \ldots\}$, where $\gcd(a, q) = 1$; see Apostol (1976, p. 148), Hardy and Wright (1979, p. 13). The asymptotic estimate for the number $\pi(x; q, a)$ of primes in the above progression that are less than or equal to $x$ is, as one would intuitively expect, $\pi(x; q, a) \approx \frac{1}{\varphi(a)} \frac{x}{\ln x}$ (Davenport 2000, p. 121). The sharper estimate $\pi(x; q, a) \approx \frac{li(x)}{\varphi(q)} + O\left(x^{1/2 + \varepsilon}\right)$ holds under a stronger assumption (the generalized Riemann hypothesis).

We can generalize Sophie Germain pairs to Cunningham chains (of the first kind), which are sequences of primes of the form $p, 2p + 1, 4p + 3, 8p + 7, \ldots, 2^{k-1}p + 2^{k-1} - 1$. For example, $2, 5, 11, 23, 47$ is such a Cunningham chain of length 5. A similar heuristic argument applies to estimating the number of primes $p$ less than or equal to $x$ such that $p$ is the first term in a first-order Cunningham chain of length $n$ as a certain constant (expressed as an infinite product over the primes) multiple of $x/(\ln x)^n$.

This can be further extended by Dickson's conjecture (Dickson 1904), which claims that given a finite set of linear forms $L_i(x) = a_i x + b_i$, $1 \le i \le n$, there are infinitely many integer values of $k$ such that $L_i(k)$ is prime for all $i$ with $1 \le i \le n$ (unless obvious existing divisibility relations prevent that). The distribution of such $k$ is assumed to obey the same estimate of the form of a constant multiple of $x/(\ln x)^k$. More generally, "Schinzel's hypothesis H" (Schinzel and Sierpinski 1958) asserts a similar result for polynomials of arbitrary degree, that is, if $f_i(x) \in \mathbb{Z}[x]$, $1 \le i \le n$, are polynomials with positive leading coefficients, then there are infinitely many integer values of $k$ such that $f_i(k)$ is prime for $1 \le i \le n$, unless an obvious condition prevents that.

The sequences (of integers, sets, or arrays) that will be discussed in this work will be positioned at the same conjecture-rich interface.

## 2.6   Some Classical Recurrent Sequences. Ducci Games

For a recurrent sequence of order $k$, the first $n$ terms (making up "the seed") $x_0, x_1, \ldots, x_{k-1}$ are given, together with a recurrence relation of the form $x_n = f(x_{n-1}, x_{n-2}, \ldots, x_{n-k})$, which holds for $n \ge k$. In MAPLE we can quickly list a desired number of terms in a recurrent sequence. For example, the first 20 terms in the "tribonacci" recurrent sequence $\{t_n\}_{n \ge 0}$ defined as $t_0 = 0, t_1 = 1, t_2 = 1, t_n = t_{n-1} + t_{n-2} + t_{n-3}$ for $n \ge 3$ can be obtained as follows:

```
> t(0):=0: t(1):=1: t(2):=1: for r from 3 to 19 do
> t(r):= t(r-1)+t(r-2)+t(r-3) end do:
> seq(t(r),r=0..19);
```

0, 0, 1, 1, 2, 4, 7, 13, 24, 44, 81, 149, 274, 504, 927, 1705, 3136, 5768, 10609, 19513, 35890

But of course, the Fibonacci sequence $(F_n)_{n \geq 0}$ (OEIS A000045) is arguably the most celebrated among the classical sequences:

$$\begin{cases} F_0 = 0, F_1 = 1, \\ F_n = F_{n-1} + F_{n-2} \text{ for } n \geq 2. \end{cases}$$

The history of the Fibonacci numbers can be traced back to ancient Sanskrit musical theory (Singh 1985). In the European space, they first appear in Fibonacci's 1202 book *Liber Abaci* [see the translation by Sigler (2002)].

MAPLE has a specific package with a built-in function for the Fibonacci numbers:

```
> with(combinat, fibonacci): seq(fibonacci(k),k=0..30);
```

0, 1, 1, 2, 3, 5, 8, 13, 21, 34, 55, 89, 144, 233, 377, 610, 987, 1597, 2584, 4181, 6765, 10946, 17711, 28657, 46368, 75025, 121393, 196418, 317811, 514229, 832040

In MATLAB we can use the Fibonacci recurrence form to produce a Fibonacci function:

```
function fibo = fibo(n)
f(1)=1;
f(2)=1;
for I=3:n;
    f(I)=f(I-1)+f(I-2);
end;
fibo=f(n);
```

The remarkable fact that the ratios of consecutive Fibonacci numbers converge to the golden section,

$$\lim_{n \to \infty} \frac{F_{n+1}}{F_n} = \frac{1 + \sqrt{5}}{2}$$

was first noticed by Johannes Kepler in his 1611 "Essay on the Six-Cornered Snowflake" (see Colin Hardie's translation (Hardie 1966).

An explicit formula for the Fibonacci numbers in terms of the golden section and its conjugate was known to De Moivre and derived by Binet (see Livio 2002, p. 108):

$$F_n = \frac{1}{\sqrt{5}}\left[\left(\frac{1+\sqrt{5}}{2}\right)^n - \left(\frac{1-\sqrt{5}}{2}\right)^n\right].$$

In connection to modular arithmetic, an interesting sequence consists of the Pisano periods or the periods of the Fibonacci numbers modulo $n$. See Mark Renault's thesis (Renault 1996) and OEIS A001175 (http://oeis.org/A001175), where the following nice MAPLE program for the sequence of Pisano periods, due to Alois P. Heinz, can be found:

```
> a:= proc(n) local f, k, l; l:= ifactors(n)[2];
>        if nops(l)<>1 then ilcm(seq(a(i[1]^i[2]), i=l))
>     else f:= [0, 1];
>           for k do f:=[f[2], f[1]+f[2] mod n];
>                    if f=[0, 1] then break fi
>             od; k
>         fi
>     end:
> seq(a(n), n=1..100);
```

    1, 3, 8, 6, 20, 24, 16, 12, 24, 60, 10, 24, 28, 48, 40, 24, 36, 24, 18, 60, 16, 30, 48, 24, 100, 84, 72,
    48, 14, 120, 30, 48, 40, 36, 80, 24, 76, 18, 56, 60, 40, 48, 88, 30, 120, 48, 32, 24, 112, 300,
    72, 84, 108, 72, 20, 48, 72, 42, 58, 120, 60, 30, 48, 96, 140, 120, 136, 36, 48, 240, 70, 24,
    148, 228, 200, 18, 80, 168, 78, 120, 216, 120, 168, 48, 180, 264, 56, 60, 44, 120, 112, 48,
    120, 96, 180, 48, 196, 336, 120, 300

Numerous Fibonacci identities can found online at Ron Knott's multimedia website on Fibonacci numbers (Knott 2016). Among them is the Cassini identity $F_{n-1}F_{n+1} - F_n^2 = (-1)^n$. Also, every issue in the journal *Fibonacci Quarterly* has a set of interesting problems in both elementary and advanced categories. It is well known that Fibonacci numbers can be expressed as diagonal sums in Pascal's triangle [for this and other patterns in Pascal's triangle see Bogomolny (2016)].

If we change the initial conditions from 0, 1 (as is the case for the Fibonacci numbers) to 2, 1, we get another interesting integer sequence: the Lucas numbers $(L_n)_{n \geq 0}$ (OEIS A000032):

$$\begin{cases} L_0 = 2, L_1 = 1, \\ L_n = L_{n-1} + L_{n-2} \text{ for } n \geq 2. \end{cases}$$

The analogue of Binet's formula is

$$L_n = \left(\frac{1+\sqrt{5}}{2}\right)^n + \left(\frac{1-\sqrt{5}}{2}\right)^n.$$

Like the Fibonacci numbers, the Lucas numbers have interesting combinatorial connections. It is fairly well known that if we assume periodic boundary conditions, the number of bit strings of length $n$ without consecutive 1's is $L_n$. Using the transfer matrix method (Gessel and Stanley 1995, Section 7), the number of such arrangements can be represented as the trace of the $n$th power of the transfer matrix $T = \begin{bmatrix} 1 & 1 \\ 1 & 0 \end{bmatrix}$ with eigenvalues $\lambda_{1,2} = (1 \pm \sqrt{5})/2$, so that $L_n = tr(T^n) = \lambda_1^n + \lambda_2^n$. Similarly, one can see that if we don't assume periodic boundary conditions, the number of bit strings of length $N$ without consecutive 1's is the Fibonacci number $F_{n+2}$.

Interesting probabilistic analogues of Fibonacci numbers have been studied (Heyde 1980; Viswanath 1999; Emb and Tref 1987).

In other types of recurrences, a vector with numerical components is expressed in terms of the previous vector $X_{k+1} = f(X_k)$ for $k \geq 0$. An interesting such recurrence is the "Ducci game," or the "$N$-number game" (Chamb and Thomas 2004), apparently originating in the late 1800s and named after Professor E. Ducci (Honsberger 1970; Furno 1981). In this kind of vector recurrence, if $X_k = (x_1, x_2, \ldots, x_{N-1}, x_N) \in \mathbb{Z}^N$, then the vector following $X_k$ will be

$$X_{k+1} = (|x_1 - x_2|, |x_2 - x_3|, \ldots, |x_{N-1} - x_N|, |x_N - x_1|).$$

It is well known that if $N$ is a power of 2, then the above iteration eventually leads to the null $N$-tuple $(0, 0, \ldots, 0, 0)$. Otherwise, the Ducci iteration may eventually enter a limit cycle of length greater than 1 with the interesting feature that for a fixed integer $C > 0$, every vector in the limit cycle has all components in the set $\{0, C\}$ (which makes the Ducci map essentially a binary iteration). Detailed data on cycle lengths of $N$-number Ducci games for $N$ up to 40 are provided in Calkins et al. (2005). The 37-number Ducci game is listed with two possible cycle lengths: 1 and 3,233,097 (the largest cycle size for the Ducci games considered in the study). To see an example in which the number of components in a Ducci vector is not a power of two, let's use MAPLE to play a "5-number game" starting from $(7, 5, 6, 7, 1)$. Note that after 10 iterations, we eventually enter a cycle of length 15. The arrays below display the preperiod and period (the 10th vector and the 25th vector coincide), with the first column playing the role of a counter for the iterative step in the Ducci game.

```
> a(1,1):=0: a(1,2):=7: a(1,3):=5: a(1,4):=6: a(1,5):=7:
a(1,6):=1: N:=26: for k from 2 by 1 to N do a(k,1):=k-1:
a(k,2):=abs(a(k-1,2)-a(k-1,3)): a(k,3):=abs(a(k-1,3)-a(k-1,4)):
a(k,4):=abs(a(k-1,4)-a(k-1,5)): a(k,5):=abs(a(k-1,5)-a(k-1,6)):
a(k,6):=abs(a(k-1,6)-a(k-1,2)): end do:
interface(rtablesize=infinity): g:=(i,j)->a(i,j): Matrix(N,6,g);
```

|   |   |   |   |   |   |   |   |   |   |   |   |   |   |   |   |   |   |   |
|---|---|---|---|---|---|---|---|---|---|---|---|---|---|---|---|---|---|---|
| 0 | 7 | 5 | 6 | 7 | 1 | 10 | 1 | 0 | 0 | 0 | 1 | 18 | 1 | 1 | 1 | 0 | 1 |
| 1 | 2 | 1 | 1 | 6 | 6 | 11 | 1 | 0 | 0 | 1 | 0 | 19 | 0 | 0 | 1 | 1 | 0 |
| 2 | 1 | 0 | 5 | 0 | 4 | 12 | 1 | 0 | 1 | 1 | 1 | 20 | 0 | 1 | 0 | 1 | 0 |
| 3 | 1 | 5 | 5 | 4 | 3 | 13 | 1 | 1 | 0 | 0 | 0 | 21 | 1 | 1 | 1 | 1 | 0 |
| 4 | 4 | 0 | 1 | 1 | 2 | 14 | 0 | 1 | 0 | 0 | 1 | 22 | 0 | 0 | 0 | 1 | 1 |
| 5 | 4 | 1 | 0 | 1 | 2 | 15 | 1 | 1 | 0 | 1 | 1 | 23 | 0 | 0 | 1 | 0 | 1 |
| 6 | 3 | 1 | 1 | 1 | 2 | 16 | 0 | 1 | 1 | 0 | 0 | 24 | 0 | 1 | 1 | 1 | 1 |
| 7 | 2 | 0 | 0 | 1 | 1 | 17 | 1 | 0 | 1 | 0 | 0 | 25 | 1 | 0 | 0 | 0 | 1 |
| 8 | 2 | 0 | 1 | 0 | 1 |   |   |   |   |   |   |    |   |   |   |   |   |
| 9 | 2 | 1 | 1 | 1 | 1 |   |   |   |   |   |   |    |   |   |   |   |   |

For the 4-number game, W.A. Webb has proved (Webb 1982) that the games in which it takes the longest time to get to the null vector are (essentially) those starting from an initial four-tuple consisting of consecutive tribonacci numbers $(t_n, t_{n-1}, t_{n-2}, t_{n-3})$, in which case the Ducci iteration takes $3\lfloor \frac{n}{2} \rfloor$ steps until it reaches the null cycle. For example, let's use MAPLE to see what happens if we start from $(t_{13}, t_{12}, t_{11}, t_{10}) = (927, 504, 274, 149)$.

| | | | | | | | | |
|---|---|---|---|---|---|---|---|---|
| 0 | 927 | 504 | 274 | 149 | 10 | 48 | 24 | 160 | 88 |
| 1 | 423 | 230 | 125 | 778 | 11 | 24 | 136 | 72 | 40 |
| 2 | 193 | 105 | 653 | 355 | 12 | 112 | 64 | 32 | 16 |
| 3 | 88 | 548 | 298 | 162 | 13 | 48 | 32 | 16 | 96 |
| 4 | 460 | 250 | 136 | 74 | 14 | 16 | 16 | 80 | 48 |
| 5 | 210 | 114 | 62 | 386 | 15 | 0 | 64 | 32 | 32 |
| 6 | 96 | 52 | 324 | 176 | 16 | 64 | 32 | 0 | 32 |
| 7 | 44 | 272 | 148 | 80 | 17 | 32 | 32 | 32 | 32 |
| 8 | 228 | 124 | 68 | 36 | 18 | 0 | 0 | 0 | 0 |
| 9 | 104 | 56 | 32 | 192 | | | | | |

For completeness let us briefly sketch here a "common-sense" proof of the fact that every four-number Ducci game eventually reaches the null state $(0, 0, 0, 0)$. Three main elementary ideas are to be considered, with the third being the "punch line" consequence of the first two.

First, the entries in the 4D vectors appearing during the Ducci iteration are bounded. If one such vector is $(a, b, c, d)$ with $0 \leq a, b, c, d \leq M$, then the entries occurring in the subsequent vector $(|a - b|, |b - c|, |c - d|, |d - a|)$ are still in the interval $[0, M]$.

Second, it is easy to see that if we reduce the Ducci iteration $(a, b, c, d) \mapsto (|a - b|, |b - c|, |c - d|, |d - a|)$ modulo 2, it becomes particularly simple. The "mod 2 Ducci" iteration is $(\alpha, \beta, \gamma, \delta) \mapsto (\alpha + \beta, \beta + \gamma, \gamma + \delta, \delta + \alpha)$. At this point, a direct verification shows that after four such modular iterations, the mod 2 Ducci iteration arrives at the null modular vector (check this!). Going back to integers, this means that an additional factor of two emerges for the entries in a vector after every block of four integer Ducci iterations. Consequently, the vectors appearing in the Ducci game are divisible by higher and higher powers of 2.

Thirdly, it is easy to see that putting together the two facts about the integer entries that "they are bounded" and "they become divisible by increasing powers of 2" leads to the conclusion that the null integer vector is eventually reached (indeed, if an integer in the interval $[0, M]$ becomes divisible by a power of 2 with $2^k > M$, it must necessarily equal 0).

The limiting binary behavior of the $N$-number Ducci iteration makes it equivalent, if we focus on the cycle only, to multiplication by $1 + x$ in the ring $F_2[x]/(x^N - 1)$. This leads to a comfortable placement of the Ducci problem in the area of cyclotomy (Calkins et al. 2005; Breuer et al. 2007). An interesting connection with Artin's conjecture is established (Breuer et al. 2007), to the effect that if $p$ is prime and 2 is a primitive root modulo $p$, then the $p$-number Ducci games have only one possible period other than 1.

Numerous other variations and generalizations of the Ducci sequences theme have been considered: analogues in higher dimensions (Breuer 2007), Ducci sequences over the reals and their asymptotic behavior (Brockman and Zerr 2007; Brown and Merzel 2003), Ducci sequences with algebraic numbers (Caragiu et al. 2011), the dynamics of Ducci sequences defined in terms of various weightings (Chamberland 2003), $p$-adic Ducci games (Car and Bax 2007), etc. A very interesting Ducci-type special iteration is introduced in Cobeli and Zaharescu (2014), where the authors play with the "atomic transformation function" $Z(a, b) = \frac{ab}{(\gcd(a,b))^2}$ (as a consequence, the authors state and prove a variation of Gilbreath's conjecture).

The Ducci game modulo 2 constitute a special case of a one-dimensional *cellular automaton* (Berto and Tagliabue 2012) encoded as "rule 102" in Wolfram's classification (Weisstein). The evolution of cellular automata reveals interesting patterns of emerging complexity and self-organization, which makes them potentially useful in the study of complex physical systems and statistical mechanics (Wolfram 1983). In the same conceptual family lies the celebrated "Game of Life" due to John Conway (Gardner 1970). Later on, we will consider a series of variations on the cellular automaton concept, with the help of special number-theoretic functions.

An infinite-dimensional Ducci game analogue, lying at the above-mentioned additive–multiplicative interface, is the classical Gilbreath conjecture in number theory (Caldwell). Invented in 1958 by napkin-doodling computer expert and amateur magician Norman O. Gilbreath (Genii Magazine), this is about successively generating a string of infinite sequences using a Ducci-type iteration starting with the sequence of primes $p_1, p_2, p_3, p_4, \ldots, p_k, p_{k+1}, p_{k+2}, \ldots$. Every sequence $x_1, x_2, x_3, x_4, \ldots, x_k, x_{k+1}, x_{k+2}, \ldots$ that occurs in the process is updated as

$$|x_1 - x_2|, |x_2 - x_3|, |x_3 - x_4|, |x_4 - x_5|, \ldots, |x_k - x_{k+1}|, |x_{k+1} - x_{k+2}|, |x_{k+2} - x_{k+3}|, \ldots.$$

Gilbreath's conjecture asserts that the first entry of each such sequence, with the exception of the first one, equals 1.

```
2  3  5  7  11  13  17  19  23  29  31  37  41  43  47  53  59  61  67  71...

1  2  2  4  2   4   2   4   6   2   6   4   2   4   6   6   2   6   4   2...

1  0  2  2  2   2   2   2   4   4   2   2   2   2   0   4   4   2   2...

1  2  0  0  0   0   0   2   0   2   0   0   0   2   4   0   2   0...

1  2  0  0  0   0   2   2   2   2   0   0   2   2   4   2   2...

1  2  0  0  0   2   0   0   0   2   0   2   0   2   2   0...

1  2  0  0  2   2   0   0   2   2   2   2   2   0   2...

1  2  0  2  0   2   0   2   0   0   0   0   2   2...

1  2  2  2  2   2   2   2   0   0   0   2   0...

1  0  0  0  0   0   0   2   0   0   2   2...

1  0  0  0  0   0   2   2   0   0   0...

1  0  0  0  0   2   0   2   0   0...
```

................................................................................

The second line in Gilbreath's table is the sequence of prime gaps (OEIS sequence of differences between consecutive primes, A001223). This brings us to the following elementary classical situation.

Once of the nicest elementary results about primes, to the effect that there are arbitrarily large gaps between consecutive primes, uses the beautiful idea of the compositeness of the terms of the arithmetic progression

$$n! + 2, \; n! + 3, \ldots, n! + n.$$

Indeed, these are $n - 1$ integers divisible by $2, 3, \ldots, n$, respectively.

As for computational evidence, in 1993 Andrew Odlyzko verified the Gilbreath conjecture up to a row rank of $3.4 \cdot 10^{11}$ (Odlyzko 1993).

## 2.7  Deeper into the Randomness

Everything we care about lies somewhere in the middle, where pattern and randomness interlace.—James Gleick, The Information: A History, a Theory, a Flood

Generating mathematical randomness is a paradox and an art. Random number generation is important for computer security, while random structures and related randomized algorithms are crucial elements in the contemporary computing environment. A great way to introduce students to higher mathematics is to let them study the $\pm 1$ sequences of Legendre symbols $\left(\frac{1}{p}\right), \left(\frac{2}{p}\right), \ldots, \left(\frac{p-1}{p}\right)$ describing the distribution of quadratic residues and nonresidues modulo a prime $p$. The study of the patterns of quadratic residues and nonresidues is an extremely interesting problem that attracted considerable attention in the period ranging from the end of the nineteenth century into the first half of the twentieth century (Aladov 1896; Davenport 1931; Davenport 1933; Peralta 1992). Ultimately, this distribution problem turns out to be intimately connected to one of the "big problems" in mathematics, the Riemann hypothesis for curves over finite fields, settled in 1949 by Weil (1949), generalizing a "genus 1" result obtained by Helmut Hasse that estimates the number of points $N$ of an elliptic curve $y^2 = x^3 + ax + b$ over the finite field with $p$ elements as $|N - (p + 1)| \leq 2\sqrt{p}$.

Students can quickly get hands-on computational experience using MAPLE. For example, if $p = 19$, the $\pm 1$ sequence of Legendre symbols can be quickly obtained as follows:

```
> with(numtheory):
> p:=19: L:=[seq(legendre(k,p),k=1..p-1)];
        L := [1, -1, -1, 1, 1, 1, 1, -1, 1, -1, 1, -1, -1, -1, -1, 1, 1, -1]
```

It is not hard (for both faculty and students) to become fascinated by this mix of global order and local randomness. Even the "smallest" problems that arise in the process are likely to have deep connections with analytic number theory. For example, just reflecting on the rank of the appearance of the first $-1$ in the $\pm 1$ sequence generated, as above, for arbitrary primes $p$ leads to the deep problem of the "least quadratic nonresidue" (Burgess 1957). The OEIS sequence listed as A053760 presents the smallest quadratic nonresidue modulo the $n$th prime for $n = 1, 2, 3, \ldots$. The first "unusually high" smallest quadratic nonresidue appears to be the one modulo $p = 311$. Indeed, using MAPLE as above, it will be easy to verify that in the sequence $\left(\frac{1}{311}\right), \left(\frac{2}{311}\right), \ldots, \left(\frac{310}{311}\right)$, the first $-1$ appears only at the eleventh position.

By incorporating MAPLE packages (in addition to **numtheory**, we will be needing **plots** and **ListTools**), it is always illuminating to visualize the sequence of partial sums $\left(\frac{1}{p}\right) + \left(\frac{2}{p}\right) + \cdots + \left(\frac{k}{p}\right)$ with $1 \le k \le p - 1$. For example, let us use $p = 1237$, a prime congruent to 1 modulo 4. This will lead us to the following globally symmetric, locally random display for the sequence of partial sums.

```
> with(numtheory): with(plots): with(ListTools):
> p:=1237; L:=[seq(legendre(k,p),k=1..p-1)]: S:=PartialSums(L):
  listplot(S);
```

$$p := 1237$$

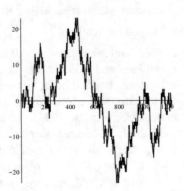

The globally symmetric shape follows from the multiplicative property of the Legendre symbol modulo the particular prime $p = 1237$, to the effect that $\left(\frac{-x}{1237}\right) = \left(\frac{-1}{1237}\right)\left(\frac{x}{1237}\right) = \left(\frac{x}{1237}\right)$. The locally random character follows, for example, from the Pólya–Vinogradov inequality (Davenport 2000), which, adapted to the case of the Legendre symbol, consists of the partial character sum estimate

$$\left| \sum_{k=N+1}^{N+H} \left(\frac{k}{p}\right) \right| \le \sqrt{p} \ln p.$$

In order to get rid of global symmetry and get plots more and more resembling one-dimensional random walks, one has the option of using Legendre symbols with a polynomial argument. For example, the use of a separable cubic polynomial $f(x) = x^3 + ax + b$ over the finite prime field $F_p$ leads us to the sets

$$X(a,b,p) = \left\{ x \in F_p \,|\, x^3 + ax + b \text{ is a square} \right\} \subseteq F_p.$$

The subsets $X(a,b,p) \subset F_p$ behave like "random" subsets with a density of (roughly) 0.5. We can define related "elliptic walks" (see Caragiu et al. 2006), where the agreement was to replace the possible zero values of the quadratic characters with 1's) with the following specifications:

$$W_{a,b,p}(k) = \begin{cases} 1, & \text{if } f(k) \text{ is a square in } F_p, \\ -1, & \text{if } f(k) \text{ is a nonsquare in } F_p, \end{cases}$$

$$B_{a,b,p}(k) = \sum_{j=1}^{k} W_{a,b,p}(j), \quad k = 1, 2, \ldots, p.$$

It was proved in Caragiu et al. (2006) through chi-square statistical testing that the above-mentioned "elliptic walks" cannot be distinguished from genuine random walks if we use criteria such as the number of returns to the origin (*the expected number of returns* to the origin for a random $\pm 1$ walk of length $2n$ returning to the origin exactly $r$ times is $\frac{1}{2^{2n-r}} \binom{2n-r}{r}$ according to (Feller 1968, p. 96), the grouping test, or the runs test (Knuth 1969, p. 74, Ex. 14).

However, if other criteria are used, such as the total displacement that for "elliptic walks" is confined (Caragiu et al. 2006) to the interval $\left[-2\sqrt{p}, 3 + 2\sqrt{p}\right]$ according to the Hasse–Weil estimate, then elliptic walks are distinguishable from genuine random walks. Indeed, for $\lambda < \mu$, the probability that the endpoint of a truly symmetric $\pm 1$ random walk of length $p$ is in the interval $\left[\lambda\sqrt{p}, \mu\sqrt{p}\right]$ equals (Feller 1968, p. 76), as a consequence of the central limit theorem, $\frac{1}{\sqrt{2\pi}} \int_{\lambda}^{\mu} e^{-\frac{x^2}{2}} dx$, so that as a consequence, a fraction of $\frac{1}{\sqrt{2\pi}} \int_{3}^{4} e^{-\frac{x^2}{2}} dx = 0.00131822\ldots$ symmetric $\pm 1$ random walks of any length $p$ have their endpoints in $\left[3\sqrt{p}, 4\sqrt{p}\right]$, a condition that is not satisfied by the elliptic walks defined as above.

In any case, a useful class exploration of (pseudo)random structures will lead students to use MAPLE to visualize the incomplete character sums (which coincide with "elliptic walks" with the few possible exceptions where $j^3 + aj + b = 0$):

$$\sum_{j=1}^{k} \left(\frac{j^3 + aj + b}{p}\right), \quad k = 1, 2, \ldots, p - 1.$$

```
> p:=8837; L:=[seq(legendre(k^3+3*k+1,p),k=1..p-1)]:
S:=PartialSums(L): listplot(S);
```

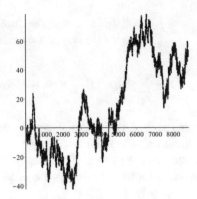

We can easily use MAPLE to visualize two-dimensional random structures. One possibility is to construct a two-dimensional analogue of elliptic walks by considering cubic polynomials over a finite field of the form $F_{p^2}$ with the prime $p$ being subject to $p \equiv 3 \pmod 4$, a constraint used for convenience due to the relation $F_{p^2} = F_p[i] = \{x + iy | x, y \in F_p\}$, which helps in visualizing $F_{p^2}$ as a $p \times p$ rectangular grid. The image below represents a projection (obtained by MATLAB) of the elliptic curve $y^2 = x^3 + x + 3$ over the base field $F_{59^2}$, that is, a pictorial description of all perfect squares in the field $F_{p^2}$.

NOTE: A more general procedure is to use a two-variable polynomial $f(x, y) \in F_p[x, y]$ with coefficients in an arbitrary finite prime field $F_p$ and display in a similar manner the points $(x, y) \in F_p \times F_p$ such that $f(x, y)$ is a perfect square in $F_p$. Using various averaging/smoothing procedures (such as replacing the value at each node of the grid with the average over a certain neighborhood), we can get similar displays in color.

Later on, we will introduce new such "models of randomness" using the greatest prime factor function and Conway's subprime function. But first, we will briefly provide an overview of functions that will play an important role in our future constructions.

## 2.8    The Greatest Prime Factor Function

For a positive integer $x$, the greatest prime factor function $P(x)$ is the largest prime factor of $x$, with the proviso that $P(1) := 1$. This will be the main function that we will be using in our sequential experiments. The OEIS entry on the greatest prime factor (A006530) was always particularly inspiring in our work on greatest prime factor sequences. It features a concise MAPLE function dealing with the greatest prime factor function due to Peter Luschny, where **factorset** (producing the set of all prime factors) is followed by *op*, which presents this set as a numerical list for which we are able to calculate the **max**:

```
> with(numtheory):
> P:=n->max(1,max(op(factorset(n))));
```

In MATLAB we can use the direct procedure

```
function gpf=gpf(n)
gpf=max(factor(n));
end
```

Over the years, the greatest prime factor function has received substantial attention from the standpoint of advanced analytic number theory, and many outstanding results and estimates have been obtained. Finding asymptotic estimates for the greatest prime factor function is an enterprise that goes a long way back. In 1930, K. Dickman investigated the probability $P(x, n)$ that a randomly selected integer $k \in \{1, 2, \ldots, n\}$ satisfies $p := P(k) < n^x$, showing that the expected value of $x$ such that the random variable $p$ is of the form $p = n^x$ is $0.62432999\ldots$, which came to be known as the *Golomb–Dickman constant* (Dickman 1930; Weisstein; Knuth and Pardo 1976).

In Kemeny (1993), it was proved that the average value of $\frac{P(n)}{n}$ is, asymptotically, $\frac{\zeta(2)}{\ln n}$, while the probability that $P(n) > \sqrt{n}$ equals $\ln 2 = 0.69314718\ldots$ in the asymptotic limit.

To list a few other examples from the fertile and engaging work on the greatest prime factor function, Erdős showed (1952) that if $f(x) \in \mathbb{Z}[x]$ is a polynomial that does not split as a product of linear polynomials in $\mathbb{Z}[x]$, and if $P_x := P\left( \prod_{k=1}^{x} f(k) \right)$, then for some $c > 0$, the estimate $P_x > x(\ln x)^{c \ln \ln \ln x}$ holds. In fact, there are numerous very interesting results on the greatest prime factor of the product of consecutive integers. In Laishram and Shorey (2005) an elementary proof is provided to the effect that $P(n(n+1)\ldots(n+k-1))$ is greater than $2k$ if $n > \max(k+13, (279/262)k)$, and it is greater than $1.97k$ if $n > k+13$.

Hooley investigated the greatest prime factor of a quadratic polynomial (Hooley 1967). Grytczuk et al. showed (2001) that the Fermat numbers $F_m = 2^{2^m} + 1$ satisfy $P(F_m) \geq 2^{m+2}(4m+9) + 1$ for $m \geq 4$. For different types of numbers, one of the results proved by Erdős and Shorey in (1976) implies that $P(2^p - 1) \gg p \ln p$ for all primes $p$. Luca and Najman (2011) provided the solution of the inequality $P(x^2 - 1) < 100$ in natural numbers.

In the previous classical intermezzo, we showed how factorials are used to build sequences of consecutive composite integers. With the greatest prime factor concept introduced, the following proposition of a similar flavor can be proved. These kinds of results at the borderline between "easy" and "wow" make for excellent tools and supplements for the undergraduate number theory classroom:

**Proposition 2.2** *Let $M > 0$ be fixed. Then for every positive integer $k$ one can find a sequence of consecutive integers $n+1, n+2, \ldots, n+k$ such that $P(n+r) > M$ for $r = 1, 2, \ldots, k$.*

*Proof* Let $p_1, p_2, \ldots, p_k$ be distinct prime numbers greater than M. With the Chinese remainder theorem (Leveque 1977, p. 60), one can infer the existence of a positive integer $n$ such that the following congruences are satisfied:

$$n \equiv -1 (\mathrm{mod} p_1)$$
$$n \equiv -2 (\mathrm{mod} p_2)$$
$$\ldots \ldots$$
$$n \equiv -k (\mathrm{mod} p_k)$$

Then for $1 \leq r \leq k$, we have $P(n+r) \geq p_r > M$, which concludes the proof.

## 2.9   Overview of Some Other Number-Theoretic Functions and Sequences

In what follows, we will briefly review some other important functions, besides the greatest prime factor, that will be used in other experiments to be conducted in our "sequential laboratory."

### (A)  The least prime factor function

The least prime factor function $p(n)$ is indexed by OEIS as A020639:

$$1, 2, 3, 2, 5, 2, 7, 2, 3, 2, 11, 2, 13, 2, 3, 2, 17, 2, 19, 2, 3, 2, 23, 2, 5, 2, 3, 2, 29, 2, 31, 2, 3.$$

In relation to the greatest prime factor $P(n)$, Erdős and van Lint (1982) proved the following asymptotic series result: $\sum_{n \leq x} \frac{p(n)}{P(n)} = \frac{x}{\ln x} + \frac{3x}{\ln^2 x}(1 + o(1))$ as $x \to \infty$.

On a truly elementary note, the next intermezzo is recognized as a constant inhabitant of virtually all undergraduate classes that involve discussions on factoring, e.g., discrete mathematics, abstract algebra, introductory number theory, fundamental mathematics for teachers.

If $n > 1$ is a composite integer, then the least prime factor of $n$ is less than or equal to $\sqrt{n}$, with equality if $n$ is the square of a prime.

The least prime factor of an integer $n$ is recorded by OEIS as A020639, where the following MAPLE routine (attributed to R.J. Mathar) is recorded:

```
> LPF:= proc(n) if n = 1 then 1; else
min(op(numtheory[factorset](n))) ; end if; end proc:
```

In MATLAB we can use the following:

```
function lf=lf(n)
lf=min(factor(n));
end
```

## (B)   Conway's subprime function

Related to the least prime factor function, Conway's "subprime function" is defined as

$$C(x) = \begin{cases} x, & \text{if } x \text{ is prime or } x = 1, \\ x/LPF(x), & \text{otherwise.} \end{cases}$$

Alternatively, $C(1) := 1$, and otherwise, $C(n)$ is the largest proper divisor of $n$. The function $C(x)$ gained prominence in relation to a class of interesting analogues of the Fibonacci recurrence, namely Conway's subprime Fibonacci sequences; see the 2012 Tatiana Khovanova blog entry (Khovanova 2012) and the 2014 *Mathematics Magazine* paper (Guy et al. 2014). More about Conway's subprime function, with related sequences and nonassociative structures, can be found in Chapter 4.

The MAPLE code used for the Conway function together with a plot of the list of the values $C(n), 1 \leq n \leq 500$ is shown below.

```
> C:=proc(n::integer) local u: if isprime(n)='true' or n=1 then
n else u:=factorset(n): n/min(seq(u[j], j=1..nops(u))) end if
end proc;
> listplot([seq(C(n),n=1..500)]);
```

Note that the first three clearly identifiable slopes in the list plot above correspond to primes, even numbers, and odd multiples of 3.

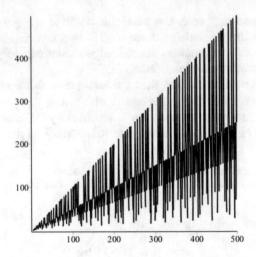

## (C)  Euler's Phi Function

We have to mention the ubiquitous Euler's totient function (see Chapter 6 in Andrews 1994), or Chapter 5 in Hardy and Wright (1979), which can be expressed in terms of the prime factors $p$ of $n$ as follows:

$$\varphi(n) = n \prod_{p|n} \left(1 - \frac{1}{p}\right).$$

This function (which can also be seen as the number of invertible elements in $\mathbb{Z}/n\mathbb{Z}$ or the number of generators of the additive group of integers modulo $n$) is one of the jewels of number theory, which has found amazing application in high-end modern cryptosystems such as RSA. A consequence of enormous importance of the totient function being the cardinality of the group of invertible elements is Euler's theorem.

**Euler's Theorem**  *For every $a$ with* $\gcd(a, n) = 1$ *we have* $a^{\varphi(n)} \equiv 1 (\mathrm{mod}\, n)$.

We will be using this theorem to build and investigate a mysterious analogue of the Fibonacci sequence.

If $p$ is a prime, then the number of primitive roots modulo $p$ is $\varphi(p - 1)$. A nice related result is the following.

**Proposition 3** *Let $\varepsilon > 0$. Then there are infinitely many primes $p$ such that the probability that a randomly chosen element in $F_p^*$ is a primitive root modulo $p$ is less than $\varepsilon$.*

*Proof* From Mertens's third theorem, we have the asymptotic relation $\prod_{p \leq x} \left(1 - \frac{1}{p}\right) \sim \frac{e^{-\gamma}}{\ln x}$ (see Section 22.8 in Hardy and Wright 1979), where $\gamma = 0.577215664\ldots$ is the Euler–Mascheroni constant. Consequently, we can find a finite set of primes $q_1, q_2, \ldots, q_k$ such that $\prod_{i=1}^{k} \left(1 - \frac{1}{q_i}\right) < \varepsilon$. Using Dirichlet's theorem, we can see that there are infinitely many primes $p$ such that $p \equiv 1 \pmod{p_1 p_2 \ldots p_k}$. For every such prime, we have $\varphi(p-1) = (p-1) \prod_{q|p-1} \left(1 - \frac{1}{q}\right) \leq (p-1) \prod_{1 \leq i \leq k} \left(1 - \frac{1}{q_i}\right) < \varepsilon(p-1)$. So the probability that a randomly chosen element in $F_p^*$ is a primitive root modulo $p$ is $\frac{\varphi(p-1)}{p-1} < \varepsilon$.

And now an interesting byproduct of this. In conjunction with a strong model-theoretic result (Chatzidakis et al. 1992), the above elementary proposition plus the fact that $\lim_{p \to \infty} \varphi(p-1) = \infty$ (in general, $\varphi(n) \geq \sqrt{\frac{n}{2}}$, Delanoy) has as a consequence that there is no formula $\Phi(x)$ in the first-order language of rings such that for every prime $p$, the set of primitive roots modulo $p$ coincides with the set of elements in the finite field $F_p$ validating $\Phi$, that is, the primitive roots are not first-order definable (Caragiu 2000).

### (D)  Lehmer's totient problem: a statistical experiment

We have seen that for a prime $p$ we have $\varphi(p) = p - 1$, but whether there are composite numbers $n$ with $\varphi(n)$ a divisor of $n - 1$ is still an open question, known as "Lehmer's totient problem." Numerous computational investigations have been made; for example, every potential solution $n$ must be greater than $10^{20}$ and have at least 14 prime divisors (Cohen and Hagis 1980), and for every potential solution $n$ that is divisible by 3, it must be the case that $n$ is greater than $10^{360,000,000}$ and has at least 40 million prime divisors (Burcsi et al. 2011).

A simple exploratory MAPLE-based exercise around the aforementioned Lehmer's problem may consider visualizing the statistical distribution of the fractional parts of the numbers $\frac{n-1}{\varphi(n)}$ in order to witness a possible unevenness that would tilt the distribution toward zero. To this end, we will visualize the histogram of the factional parts $\left\{\frac{n-1}{\varphi(n)}\right\}_{n=1}^{1,000,000}$. Here is what we get:

```
> with(numtheory): with(Statistics):
> L:=[seq(evalf(frac((n-1)/phi(n))),n=1..1000000)]: Histogram(L,
frequencyscale=absolute, bincount=100);
```

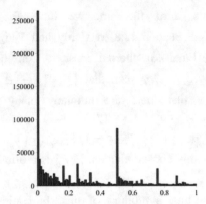

These computer algebra visuals have the potential of becoming pivotal educational moments that will inspire undergraduates to become involved with further research in number theory: in their journey, this could be the first step. Notice a major "accumulation point" toward zero and a minor accumulation point toward the middle, 0.5.

### (E)  The $3x + 1$ (Collatz) problem

Erdős offered \$500 for the solution to this problem, while offering the sobering assessment to the effect that "*Mathematics may not be ready for such problems*" (Lagarias 1985). It states that for every seed $x_0$, the sequence defined by the recurrence

$$x_{n+1} = \begin{cases} 3x_n + 1, & \text{if } x_n \text{ is odd,} \\ \frac{x_n}{2}, & \text{if } x_n \text{ is even.} \end{cases}$$

eventually reaches 1 (and hence enters the cycle 1, 4, 2). Serious computing power has been invested in the problem, with many articles published featuring generalizations of Collatz's problem. When it comes to formal results, Krasikov and Lagarias (2003) proved that for all sufficiently large values of $x$, at least $x^{0.84}$ integers $S$ from 1 to $x$ have the property that the Collatz iteration with seed $x_0 = s$ eventually reaches 1. Collatz's problem has many connections. Particularly interesting are the those (Kontorovich and Miller 2005; Lagarias and Soundararajan 2006) with Benford's law (Hill 1996) for the distribution of the leading digit in numerous real-life data sets.

An unusually long preperiod in Collatz's iteration occurs for $x_0 = 27$ (see OEIS's A008884, $3x + 1$ sequence starting at 27). The following MAPLE set of instructions (Monagan) can be used to obtain a view on the process, a bumpy ride toward the limit cycle.

```
> with(plots):
> x[0]:= 27;
> for i from 0 to 200 do:
> if x[i] mod 2=0 then x[i+1] := x[i]/2:
> else x[i+1] := 3*x[i]+1 fi;
> od;
> L:=[seq(x[i],i=1..200)]: listplot(L);
```

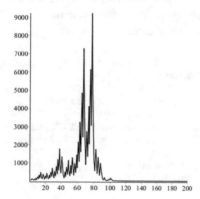

In MATLAB, this can be achieved by

```
function collatz = collatz(a,n)
x(1)=a;
for I=2:n;
    if mod(x(I-1),2)==0;
        x(I)=x(I-1)/2;
    else
        x(I)=3*x(I-1)+1;
    end
end
collatz=plot(x);
```

The On-Line Encyclopedia of Integer Sequences has a wealth of entries (714 as of November 2016) in reference to Collatz's problem, with several involving prime numbers, e.g., A078350 (the number of primes in the Collatz iteration starting with $n$), A070975 (the number of steps required to get to 1 if one starts from the $n$th prime), etc.

## 2.10   MATLAB Too!

We have been speaking mostly about MAPLE, but MATLAB is equally useful: we love them both! However, the author and his students at the time found custom-made functions somehow easier to write in MATLAB (although probably if we had known more about MAPLE, that opinion would have changed). We would like to share a few of these functions that were either used or attempted in senior capstone projects.

- **The Blum–Blum–Shub random number generator**

This is a random number generator (see Blum et al. 1986) of importance to cryptography, which uses a modulus $M = pq$ with $p, q$ large primes. It calculates terms $x_i$ using the recurrence $x_{n+1} = x_n^2 (\mathrm{mod}\, N)$ from an initial "seed" $x_0 = a$, and at every step takes $x_i \bmod 2 \in \{0, 1\}$ as a term in the generated pseudorandom bit string. For the purpose of improved visualization, we will instead work with the associated $\pm 1$ strings (just use the function $x \mapsto 2x - 1$ to the effect $0 \mapsto -1, 1 \mapsto 1$) and plot their cumulative sums.

```
function bbs = bbs(p,q,n,a)
m=p*q;
x(1)=a
z(1)=2*mod(a,2)-1;
for I=2:n;
    x(I)=mod(x(I-1)^2,m);
    z(I)=2*mod(x(I),2)-1;
end
y=cumsum(z);
bbs=plot(y)
```

For example, the choice $p = 983, q = 967$ with $a(= x_0) = 103$ and $N = 1000$ steps will produce the following image for the associated $\pm 1$ walk:

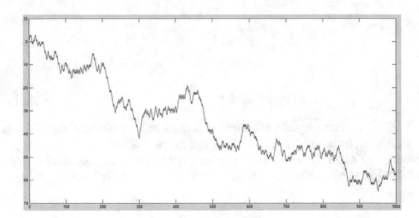

> **Functions related to the Euclidean algorithm**

The following MATLAB function calculates the number of divisions in the Euclidean algorithm (also see OEIS entry A051010)

```
function euclid = euclid(a,b)
h=max(a,b);
l=min(a,b);
I=0;
while l ~= 0;
    I=I+1;
    r=mod(h,l);
    h=l;
    l=r;
end;
euclid=I;
```

Lamé's theorem asserts that the number of divisions in the Euclidean algorithm is at most five times the number of digits of the smaller number, with the maximum attained for pairs of consecutive Fibonacci numbers. If you want to get a nice color plot displaying these numbers over an $n \times 1$ grid, use the following function.

```
function euclidmat = euclidmat(n)
for I=1:n;
    for J=1:n;
        x(I,J)=euclid(I,J);
    end;
end;
euclidmat=pcolor(x)
```

For example, **euclidmat(1000)** will produce the following display:

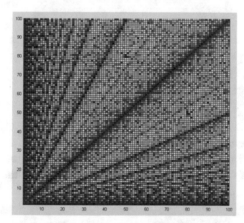

As for the obviously visible edges in the image, the "valleys" (minima) corresponds to the pairs of numbers that are multiples of each other, while the "ridges" (maxima) are related (as their slopes in the image) to the quotients of consecutive Fibonacci numbers.

The next function produces a plot displaying the evolution (exponential decay) of the sequence of successive remainders in a particular Euclidean algorithm.

```
function euclidgraph = euclidgraph(a,b)
h=max(a,b);
l=min(a,b);
I=0;
while l ~= 0;
    I=I+1;
    x(I)=l;
    r=mod(h,l);
    h=l;
    l=r;
end;
euclidgraph=plot(x);
```

## ➢ Pollard's rho

Randomized factoring using the polynomial $x^2 + a$; note that students are encouraged to experiment with their "favorite" functions)

```
function rhotest = rhotest(n,a)
x=2;
y=2;
d=1;
J=0;
while d==1;
    x=rem(x^2+a,n);
    y=rem((y^2+a)^2+a,n);
    d=gcd(abs(x-y),n);
    J=J+1;
end
if d==n;
    rhotest=0
else
    rhotest=d;
    J
end
```

## ➢ Prime walks

Choose a sequence of odd primes $3, 5, 7, 11, 13, 17, \ldots$. To each term we associate $+1$ or $-1$ depending on whether the prime is congruent to 1 or 3 modulo 4. The sequence of cumulative sums will give us a view of the interwoven distribution of primes in the two congruence classes modulo 4.

```
function [primewalk]=primewalk(m)
x=primes(m);
k=size(x,2);
for I=2:k;
    x(I)=2-mod(x(I),4);
end;
y=cumsum(x);
primewalk=plot(y);
```

For example, **primewalk(100000)** will produce the following:

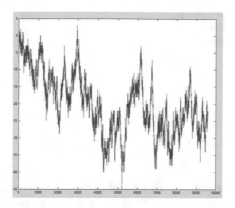

## ➤ A computational exercise with a "Collatz-type" generation

The following is a problem in the category of "startups," that is, new problems that are interesting enough that we may conjecture that they will attract some attention in the future. That said, let us consider the following function inspired by Collatz's recurrence:

$$Coll(x,y) = \begin{cases} (x+y)/2 & \text{if } x \equiv y(\text{mod } 2), \\ 3(x+y)+1, & \text{otherwise.} \end{cases}$$

It may be seen as a binary operation on the set of natural numbers. This MATLAB exercise will engage students in the exploration of the structure generated by an initial "seed set" $A$ by recursively applying the Collatz operation starting from $X_0 = A$. We are thus looking at the set recurrence

$$X_{n+1} = X_n \cup \{Coll(x,y)|x,y \in X_n\}.$$

The MATLAB function for *Coll* is

```
function coll = coll(x,y)
if mod(x+y,2)==0;
        coll=(x+y)/2;
    else
        coll=3*x+3*y+1;
    end
end
```

The set generation will be performed by the following function:

```
function [groupoidcollatz] =
groupoidcollatz(x)
N=size(x,2);
K=N;
y=x;
for I=1:N;
    for J=1:N;
        a=coll(I,J);
        K=K+1;
        y(K)=a;
    end
end
groupoidcollatz=sort(distinct(y));
end
```

So let's begin! Note that if we start with $A = \{1\}$, then since $Coll(1,1) = 1$, all the subsequent iterates will equal $\{1\}$. Let us work then with the seed $X_0 = A = \{2\}$. In this case, the first iteration leads to $X_1 = \{2,7\}$:

>> groupoidcollatz(2)

ans =

2   7

The second iteration leads to $X_2 = \{2,7,17,29\}$, the third iteration leads to $X_3 = \{2,7,17,19,23,29,31,37,59,61,79,89,137,167\}$, and the fourth iteration gives an 85-element set:

$X_4$ = {2, 5, 7, 11, 13, 17, 19, 23, 29, 31, 37, 41, 43, 47, 53, 59, 61, 67, 71, 73, 79, 83, 89, 97, 101, 107, 113, 127, 131, 137, 139, 149, 151, 157, 163, 167, 173, 179, 181, 191, 193, 199, 211, 227, 241, 251, 257, 269, 277, 293, 307, 313, 331, 337, 347, 353, 359, 367, 373, 383, 409, 419, 421, 439, 443, 449, 467, 479, 487, 499, 503, 509, 569, 577, 593, 599, 617, 641, 647, 659, 719, 727, 761, 857, 1049}.

This will be rapidly followed by higher-order generating sets with cardinalities increasing very rapidly:

$$|X_5| = 616, |X_6| = 3875, |X_7| = 24,061,\ldots$$

Clearly, the set of natural numbers cannot be covered in this way, since, for example, the element 1 cannot be written as $Coll(x,y)$ with $x,y \in \mathbb{N}$ appearing at a previous level unless $x = y = 1$.

A myriad of questions opens for exploration at this point, for the excitement of both students and faculty participating in the project. What happens, then, if we change the seed set to $A = \{1,2\}$? To $A = \{1,2,4\}$? Can we find instances with a reasonable description of $\bigcup_{n=0}^{\infty} X_n$? Which primes will eventually fall into one of the $X_n$? The sky is the limit!

## 2.11 An Experiment with Pairs of Primitive Roots Modulo Primes

In classical number theory, a great deal of attention has been paid to certain special pairs of primitive roots modulo a prime number, for which either

- the sum is 1; we shall call these "Category 1" pairs $(\alpha, \beta)$ of primitive roots modulo $p$ such that $\alpha + \beta = 1$, or
- the difference is 1; we hall call these "Category 2" pairs $(\alpha, \alpha + 1)$ of primitive roots.

We can use MAPLE to quickly find the least primitive root modulo a prime $p$, traditionally denoted by $g_p$. The sequence of least primitive roots modulo the $n$th prime is listed in OEIS as A001918. MAPLE can quickly produce an initial segment:

```
> gp:=n->pprimroot(ithprime(n));
> seq(gp(n),n=1..100);
    1, 2, 2, 3, 2, 2, 3, 2, 5, 2, 3, 2, 6, 3, 5, 2, 2, 2, 2, 7, 5, 3, 2, 3, 5, 2, 5, 2, 6, 3, 3, 2, 3, 2, 2, 6, 5, 2, 5, 2, 2,
    2, 19, 5, 2, 3, 2, 3, 2, 6, 3, 7, 7, 6, 3, 5, 2, 6, 5, 3, 3, 2, 5, 17, 10, 2, 3, 10, 2, 2, 3, 7, 6, 2, 2, 5, 2,
    5, 3, 21, 2, 2, 7, 5, 15, 2, 3, 13, 2, 3, 2, 13, 3, 2, 7, 5, 2, 3, 2, 2
```

Grosswald provided an estimate for the size of the least primitive root, proving that $g_p < p^{0.499}$ for all large enough $p$ (Grosswald 1981). As an important existence result, it has been proved (Moreno and Sotero 1990) that if $q > 2$, then the finite field $F_q$ contains two (not necessarily distinct) primitive elements $\alpha, \beta$ such that $\alpha + \beta = 1$.

Note that there is a direct connection between the Category 1 pairs of primitive roots and the so-called "Costas arrays" (the "Golomb construction") with applications to radar and sonar systems (Costas 1984; Cohen-Mullen 1991).

Regarding Category 2 pairs of primitive roots, a conditional result, due to Vegh, shows that if $p$ is a prime greater than 3 such that $\varphi(p)/(p-1) > 1/3$, then there exists at least one Category 2 pair of primitive roots modulo $p$ (Vegh 1968). Later, S.D. Cohen showed (Cohen 1985) unconditionally that every finite field $F_q$ with $q > 3$, $q \not\equiv 7 \pmod{12}$, $q \not\equiv 1 \pmod{60}$ contains a pair of consecutive primitive roots.

It is easy to see that there are Category 2 pairs modulo 2, 3, and 7. If $\mathbf{C}$ is the set of prime powers $q$ such that there exists a pair of primitive roots of the form $\alpha, \alpha + 1$ in the finite field $F_q$, the following is conjectured by S.D. Cohen in Cohen (1985):

**Conjecture** (Cohen 1985) *The set $\mathbf{C}$ contains all prime powers except 2, 3, and 7.*

We will now experiment a little with "small" special pairs of primitive roots. First of all, let us agree on the following definition of "smallness" for such pairs: for Category 1 pairs, $\alpha, \beta$ (primitive roots with sum 1, where $\alpha \leq \beta$), we will define "small" to signify that $\alpha < \sqrt{p}$. That is, they are pairs of the form $k, p + 1 - k$ with $k < \sqrt{p}$. For Category 2 pairs $\alpha, \alpha + 1$ (primitive roots with difference 1), we will again define "small" to signify that $\alpha < \sqrt{p}$.

A conjecture regarding these "small pairs" was formulated together with my student J.C. Schroeder in the context of his senior research (Schroeder 2013), done in 2013:

**(CSPPR) Conjecture on Small Special Pairs of Primitive Roots** For all large enough primes $p$, one can find small special pairs of primitive roots of Category 1 (with sum 1) as well as small special pairs of primitive roots of Category 2 (with difference 1).

Finding computational support for the CSPPR conjecture was exciting, especially the work on a "lab manual" showing the data collection performed during that period.

A preliminary analysis involves the first 16 primes. In these cases, the existence of such small pairs generally does not hold, with the exception of the prime 29, which allows for both the Category 1 pair (3, 27) and the Category 2 pair (2, 3) Table 2.1.

As we continue, we notice that gradually more and more cases appear in which both Category 1 and Category 2 special pairs of primitive roots do exist.

**Table 2.1** Very small primes

| Prime $p$ | Category 1 (sum 1) Small (least $< \sqrt{p}$) | Category 2 (difference 1) Small (least $< \sqrt{p}$) |
|---|---|---|
| 2 | no | no |
| 3 | no | no |
| 5 | no | (2, 3) |
| 7 | no | no |
| 11 | no | no |
| 13 | no | no |
| 17 | no | no |
| 19 | no | (2, 3) |
| 23 | no | no |
| 29 | *(3, 27)* | *(2, 3)* |
| 31 | no | no |
| 37 | no | no |
| 41 | no | no |
| 43 | no | no |
| 47 | *(5, 43)* | no |
| 53 | *(3, 51)* | *(2, 3)* |

| Prime $p$ | Category 1 | Category 2 |
|---|---|---|
| *59* | *(6, 54)* | *no* |
| 61 | (7, 55) | (6, 7) |
| *67* | *(7, 61)* | *no* |
| *71* | *(7, 65)* | *no* |
| *73* | *no* | *no* |
| 79 | (3, 77) | (6, 7) |
| 83 | (5, 79) | (5, 6) |
| 89 | (7, 83) | (6, 7) |
| *97* | *no* | *no* |
| 101 | (3, 99) | (2, 3) |
| 103 | (5, 99) | (5, 6) |
| 107 | (5, 103) | (5, 6) |
| *109* | *no* | *(10, 11)* |
| 113 | (6, 108) | (5, 6) |
| *127* | *no* | *(6, 7)* |
| *131* | *(6, 126)* | *no* |
| 137 | (6, 132) | (5, 6) |
| *139* | *no* | *(2, 3)* |
| 149 | (3, 147) | (2, 3) |
| 151 | (6, 146) | (6, 7) |
| 157 | (6, 152) | (5, 6) |
| 163 | (11, 153) | (11, 12) |
| *167* | *(5, 163)* | *no* |
| 173 | (3, 171) | (2, 3) |
| 179 | (6, 174) | (6, 7) |
| *181* | *no* | *no* |
| *191* | *no* | *no* |
| *193* | *no* | *no* |
| 197 | (3, 195) | (2, 3) |
| *199* | *(3, 197)* | *no* |
| 211 | (7, 205) | (2, 3) |
| 223 | (10, 214) | (5, 6) |
| 227 | (5, 223) | (13, 14) |
| 229 | (7, 223) | (6, 7) |
| 233 | (6, 228) | (5, 6) |

The trend towards fewer instances of "no" continues; here are the data for the primes from 1967 to 1999, the last double "no" appearing for the prime 1873:

| Prime $p$ | Category 1 | Category 2 |
|-----------|------------|------------|
| 1867 | (12, 1856) | (12, 13) |
| 1871 | (14, 1858) | (41, 42) |
| *1873* | *no* | *no* |
| 1877 | (12, 1866) | (11, 12) |
| 1879 | (19, 1861) | (11, 12) |
| 1889 | (7, 1883) | (6, 7) |
| 1901 | (3, 1899) | (2, 3) |
| 1907 | (5, 1903) | (5, 6) |
| 1913 | (6, 1908) | (5, 6) |
| 1931 | (8, 1924) | (13, 14) |
| 1933 | (15, 1919) | (14, 15) |
| 1949 | (3, 1947) | (2, 3) |
| 1951 | (3, 1949) | (41, 42) |
| 1973 | (3, 1971) | (2, 3) |
| 1979 | (6, 1974) | (17, 18) |
| 1987 | (5, 1983) | (2, 3) |
| 1993 | (21, 1973) | (20, 21) |
| 1997 | (3, 1995) | (2, 3) |
| 1999 | (6, 1994 | (29, 30) |

We notice progressively a larger density of primes with small special pairs of primitive roots of both categories. Indeed, every prime starting from the 2362nd to the 10,000th displays both Category 1 and Category 2 small pairs of primitive roots. To see this, the approach was to introduce, for every prime $p$, the functions

$$\min\{2p - 2 - ord(p - k + 1) - ord(k)\}_{k=2..\lfloor\sqrt{p}\rfloor}$$

and

$$\min\{2p - 2 - ord(k) - ord(k + 1)\}_{k=2..\lfloor\sqrt{p}\rfloor}.$$

Note that these are both zero (or equivalently, their sum is zero) precisely if $p$ has a small Category 1 pair of primitive roots and also a small Category 2 pair of primitive roots. Consequently, the MAPLE function

```
> S:=r->min(seq(2*ithprime(r)-2-order(ithprime(r)-
k+1,ithprime(r))-
order(k,ithprime(r)),k=2..floor(sqrt(ithprime(r)))))+min(seq(2*i
thprime(r)-2-order(k,ithprime(r))-
order(k+1,ithprime(r)),k=1..floor(sqrt(ithprime(r)))));
```

is zero at $r$ precisely when the $r$th prime allows for both types of small pairs of primitive roots. A plot of the list of the values $S(r)$ will reveal the "zero regions" corresponding to the primes with both types of small pairs of primitive roots, thus allowing us to lean toward the plausibility of the CSPPR conjecture.

```
> L:=[seq(S(r),r=5..10000)]: listplot(L);
```

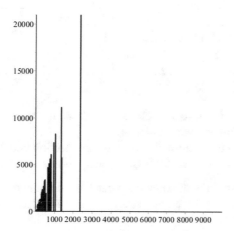

It is conceivable that the CSPPR conjecture follows in even stronger forms from far-reaching hypotheses like the generalized Riemann hypothesis, which would imply a "power-of-the-logarithm" $O(\ln^C p)$ upper bound for the least primitive root modulo $p$.

## 2.12 Traffic Flow and Quadratic Residues

Traffic flow theory began to emerge as a sophisticated area of application of mathematics in the second half of the twentieth century (Mannering and Kilareski 1990; May 1990). It is a huge area, but we will be concerned with only a particularly simple cellular automaton (CA) (Weisstein) model for one-dimensional traffic flow, the CA "rule 184" or "traffic rule" (Rosenblueth and Gershenson 2011; Wentian 1987). This CA model was the topic of a senior research course in 2010 at Ohio Northern University (Brace 2010). Here we plan to use it to visualize what happens if perfect squares modulo a prime $p$ are assimilated to "vehicles" riding on an imaginary "highway."

Under rule 184, a CA iteration consists in making any cell hosting a 1 advance one step to the right, provided the cell sitting in front of it hosts a 0. If we have only one cell hosting a 1 ("the vehicle") and otherwise only zeros, then the 1 will continue moving at a constant speed to the right. In general, we have more vehicles on the highway. Say we have 10 cells (with periodic boundary conditions and that begin moving rightward) from the following initial state:

| 1 | 1 | 0 | 1 | 0 | 0 | 1 | 0 | 1 | 0 |
|---|---|---|---|---|---|---|---|---|---|

Then the next couple of states will be as follows:

| 1 | 0 | 1 | 0 | 1 | 0 | 0 | 1 | 0 | 1 |
|---|---|---|---|---|---|---|---|---|---|
| 0 | 1 | 0 | 1 | 0 | 1 | 0 | 0 | 1 | 1 |
| 1 | 0 | 1 | 0 | 1 | 0 | 1 | 0 | 1 | 0 |
| 0 | 1 | 0 | 1 | 0 | 1 | 0 | 1 | 0 | 1 |
| 1 | 0 | 1 | 0 | 1 | 0 | 1 | 0 | 1 | 0 |

We see that this leads to a cycle of length 2. In general, though, the traffic patterns are very complex and nonlinear.

As a nice elementary exercise illustrating a connection with Fibonacci and Lucas numbers, it is easy to see that for a CA device under rule 184 with $n$ cells:

- The number of states with no blocked car ("maximum speed" configurations) is $L_n$ (example: for $n = 4$ we have seven such configurations: 1010, 0101, 1000, 0100, 0010, 0001, 0000).
- The number of states with exactly one blocked car is $nF_{n-2}$ (example: for $n = 5$, we have ten such configurations: 11000, 01100, 00110, 00011, 10001, 11010, 01101, 10110, 01011, 10101).

As a consequence, for such a CA model of traffic flow, the probability that a random distribution of vehicles "flowing freely" (i.e., no car is blocked) is

$$L_n/2^n \approx (\alpha/2)^n.$$

In general, one can prove the following generating function relation for the numbers $v(n, k, l)$ of CA states in an $n$-cell device evolving under Rule 184, with $k$ cars (1's) such that $l$ of them are able to move under the 184 updating process:

$$\left( \frac{q + 1 + \sqrt{(q-1)^2 + 4qz}}{2} \right)^n + \left( \frac{q + 1 - \sqrt{(q-1)^2 + 4qz}}{2} \right)^n$$
$$= \sum_{k,l} v(n, k, l) q^k z^l.$$

Note that we may use the quantity $l/k$ as a measure of the "average speed" in traffic (that is, the proportion of cars able to move).

Here is a list of selected MATLAB programs used in the aforementioned capstone project, instrumental in obtaining visuals on the density–speed relation and the traffic flow dynamics on an $n$-cell model, the intent being to witness, especially near the critical point where the overall car density is given by $k/n \approx 1/2$, cluster formation or shock wave propagation (going forward or backward depending on the vehicle density in a given area). Note that such issues, especially in the "thermodynamic limit" (large lattice size) indicate bridges between traffic flow theory and statistical physics (Fukś and Boccara 1998; Fukui and Ishibasi 1996).

**ardm(p)** This function generates a 0 or a 1 with the probability of a 1 being $p$.

```
function ardm=ardm(p)
x=rand(1);
if(x<p)
ardm=1;
else
ardm=0;
end
```

**cell(a,b,c)** This function represents the local neighborhood iteration according to Rule 184.

```
function cell = cell(a,b,c)
if 4*a+2*b+c==0
   cell=0;
elseif 4*a+2*b+c==1;
   cell=0;
elseif 4*a+2*b+c==2;
   cell=0;
elseif 4*a+2*b+c==3;
   cell=1;
elseif 4*a+2*b+c==4;
   cell=1;
elseif 4*a+2*b+c==5;
   cell=1;
elseif 4*a+2*b+c==6;
   cell=0;
else
   cell=1;
end
```

**next(x)** This function receives as input the bit vector $x$ and produces a vector by applying Rule 184 to every local neighborhood of three cells.

```
function next = next(x)
n=size(x,2);
a(1)=cell(x(n),x(1),x(2));
a(n)=cell(x(n-1),x(n),x(1));
for I=2:n-1;
```

```
    a(I)=cell(x(I-1),x(I),x(I+1));
end;
next=a;
```

**nextflow(x)** This function receives as input a certain highway state $x$ and calculates the average highway speed as the quotient of the number of vehicles that are actually moving and the total number of vehicles on the highway.

```
function nextflow = nextflow(x)
n=size(x,2);
d=0;
for I=1:n;
   d=d+x(I);
end;
for I=1:n-1;
   b(I)=x(I)-(x(I)*x(I+1));
   b(n)=x(n)-x(n)*x(1);
end;
c=0;
for I=1:n;
   c=c+b(I);
end;
nextflow=c/max(d,1);
```

**randvp(prob,n)** This function generates a random bit distribution on $n$ cells, where 1's appear with the probability *prob*.

```
function randvp = randvp(prob,n)
for I=1:n;
   x(I)=randp(prob);
end;
randvp=x;
```

**spdensity(n)** This function produces a plot relating the car density and average car speed.

```
function spdensity = spdensity(n)
for I=1:n;
   x=randvp(I/n,n);
   y(I)=nextflow(x);
end;
spdensity=plot(y);
```

**rflow(n,c,k)** This function displays a visual of the traffic flow dynamics on an *n*-cell model, with an initial car distribution of density *c*, analyzed for the first *k* units of simulation time.

```
function rflow = rflow(n,c,k)
x=randvp(c,n);
a(1,:)=x;
for I=2:k;
    a(I,:)=next(a(I-1,:));
end
flow=imagesc(a);
```

For example, the following image is the output of **rflow(50, 0.4, 40)**, that is, we begin with a random distribution of "vehicles" (1's in the automaton) with projected density 0.4, and run it for 40 steps. Hints of backward-moving waves can be noticed at the beginning (0.4 is not too far from the critical point 0.5), after which the movement becomes uniform, with each vehicle having an available empty cell to the right:

The following is the output of **spdensity(1000)**, illustrating through simulation the linear negative correlation between average vehicle speed and vehicle density:

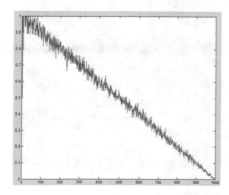

Inspired by the above traffic flow simulations, we find it exciting to try to apply the CA rule 184 to an initial structure consisting of a bit string of length $p - 1$ (where $p$ is a prime) with periodic boundary conditions, where the 1's indicate the locations of squares mod $p$, while the zeros indicate the locations of the nonsquares. In other words, the "vehicles" are the quadratic residues mod $p$. The following function creates a binary sequence of length $p - 1$ obtained from the $\pm 1$ sequence of Legendre symbols modulo $p$ by replacing the $-1$ values with 0.

```
function quadseq = quadseq(p)
for I = 1:p-1;
    x(I)=rem(I^2,p);
end
y=sort(x);
z=unique(y);
for I=1:p-1;
    s(I)=-1;
end
k=(p-1)/2;
for I=1:k
    s(z(I))=1;
end;
quadseq=s;
```

If we take this to be the initial "vehicle distribution" under the traffic CA rule, we will use the following function:

```
function quadflow = quadflow(p,k)
x=quadseq(p);
a(1,:)=x;
for I=2:k;
    a(I,:)=next(a(I-1,:));
end
quadflow=imagesc(a);
```

Let's visualize what happens with, say, the small prime 19 if we perform 20 iterations:

Note that after an initial turbulence with a compact small group of vehicles briefly "flowing backward," the circulating residues attain an equilibrium: a circular arrangement of 18 cells with periodic boundary conditions, in which every 1 is followed by a 0.

A similar phenomenon occurs if we try the prime 41. Again, the equilibrium is quickly attained after an initial traffic jam, with the 20 vehicle residues eventually traveling, equally spaced, at maximum speed:

An interesting question arises: is it true that if CA rule 184 is applied to the standard initial configuration of 1's and 0's corresponding to the quadratic residues and nonresidues modulo a prime, the configuration always stabilizes into one of equally spaced vehicles? For example, in the above scenario, the quadratic residues traffic stabilizes at the 20th step.

Here are some MATLAB functions used:

**npsum(x)** calculates the sum of the products of the nearest neighbors in a vector **x** (so that, particularly for the case of a binary "traffic vector" **x**, the value of **npsum (x)** is zero precisely when the traffic is free).

```
function [npsum]=npsum(x)
n=size(x,2);
for I=1:n-1;
  z(I)=x(I)*x(I+1);
end;
npsum=x(n)*x(1)+sum(z);
```

**quadflowend(p,k)** calculates the sums of the products of nearest neighbors for all subsequent vectors in the traffic iteration with $k$ steps that starts with the binary vector of quadratic residues modulo $p$ and places them in an output vector:

```
function quadflowend = quadflowend(p,k)
x=quadseq(p);
a(1,:)=x;
z(1)=npsum(x);
for I=2:k;
   a(I,:)=next(a(I-1,:));
   z(I)=npsum(a(I-1,:));
end;
quadflowend=z;
```

**testqfe(x)** calculates the level least $k$ such that the components of $x$ satisfy $x_i = 0$ for all $i \geq k$:

```
function [testqfe] = testqfe(x)
testqfe=size(x,2);
I=testqfe
while x(I)==0
   testqfe=testqfe-1;
   I=I-1;
end;
```

The following table indicates the level of stabilization for the primes from 7 to 100.

| $p$ | $L$ | $p$ | $L$ |
|-----|-----|-----|-----|
| 7   | 2   | 47  | 18  |
| 11  | 5   | 53  | 26  |
| 13  | 6   | 59  | 29  |
| 17  | 7   | 61  | 27  |
| 19  | 7   | 67  | 26  |
| 23  | 9   | 71  | 20  |
| 29  | 13  | 73  | 36  |
| 31  | 10  | 79  | 26  |
| 37  | 18  | 83  | 41  |
| 41  | 20  | 89  | 44  |
| 43  | 17  | 97  | 48  |

For an upper bound, it appears that a uniform flow is attained in fewer than $\frac{p-1}{2}$ iterations. Here is a table with the number of iterations to stabilization for the segment of primes between 809 and 997:

| p   | L   | p   | L   |
|-----|-----|-----|-----|
| 809 | 401 | 907 | 362 |
| 811 | 322 | 911 | 260 |
| 821 | 385 | 919 | 306 |

(continued)

(continued)

| p | L | p | L |
|---|---|---|---|
| 823 | 290 | 929 | 463 |
| 827 | 379 | 937 | 468 |
| 829 | 411 | 941 | 470 |
| 839 | 381 | 947 | 437 |
| 853 | 426 | 953 | 465 |
| 857 | 427 | 967 | 293 |
| 859 | 325 | 971 | 485 |
| 863 | 339 | 977 | 418 |
| 877 | 438 | 983 | 393 |
| 881 | 440 | 991 | 330 |
| 883 | 353 | 997 | 498 |
| 887 | 354 | | |

The next image displays 600 iterations of the traffic map starting with the bit string corresponding to the quadratic residues and nonresidues modulo 997:

There is also the question of finding a lower bound for the number of iterations needed to stabilize the traffic. We can relate this to the sequence OEIS A048280 (the length of the longest run of consecutive quadratic residues modulo the $n$th prime) by noticing that a run of $K$ consecutive 1's would need (if the environment allows, so at best) $K - 1$ traffic-rule iterations in order to separate the 1's. Thus the largest residue run minus one makes for an obvious lower bound. Of course, there is still plenty of large scale computational work to do in this direction: as in all examples in this book, the reader is invited to join in the fun.

In the "further work" category, note that we can conceivably imagine plenty of other potentially interesting iterations starting from the binary quadratic residue seed. For example, imagine we want to apply a series of Ducci iterations. This time, the iteration does not preserve (as does CA rule 185) the number of 1's. Exploring the sequence of the periods of the binary Ducci iterations starting with these particular seeds may be an interesting task.

Since we will be working with binary vectors only, we will use the following MAPLE procedure for the Ducci iteration step:

```
> Ducci:= proc(x::list)::list;
> local n, k, X;
> n:=numelems(x);
> for k from 1 to n-1 do X[k]:=(x[k]+x[k+1]) mod 2 end do;
> X[n]:=(x[n]+x[1]) mod 2;
> [seq(X[k],k=1..n)];
> end proc;
```

When it comes to finding the limit cycle lengths, we will use the tortoise and the hare (Floyd's) cycle-finding approach outlined in Section 2.4. The Ducci period corresponding to the quadratic residue seed associated to the prime $p$, say $p = 103$, is 510, and it can be obtained as follows:

```
    p:=103: for k from 1 to p-1 do s(k):=(1+legendre(k,p))/2:
end do: L:=[seq(s(k),k=1..p-1)]: SLOW:=Ducci(L):
FAST:=Ducci(Ducci(L)): k:=1: while SLOW<>FAST do
SLOW:=Ducci(SLOW): FAST:=Ducci(Ducci(FAST)): k:=k+1; end do:
TENTATIVE_START:=k: x:=SLOW: SLOW:=Ducci(x):
FAST:=Ducci(Ducci(x)): k:=1: while SLOW<>FAST do
SLOW:=Ducci(SLOW): FAST:=Ducci(Ducci(FAST)): k:=k+1; end do:
PERIOD:=k;
```

The calculations of Ducci periods thus defined vary greatly in both output value and the running time of the cycle-finding algorithm (which is conceivable if we think about the significant differences between the multiplicative structures of the numbers $p - 1$. While a pattern is not clearly detectable to us, we include a table with a couple of the periods associated to the primes from 3 to 103; the largest entry, 950,214, corresponds to the prime 59.

| $p$ | period | $p$ | period |
|-----|--------|-----|--------|
| 3   | 1      | 47  | 4094   |
| 5   | 1      | 53  | 3276   |
| 7   | 6      | 59  | 950,214 |
| 11  | 30     | 61  | 60     |
| 13  | 12     | 67  | 2046   |
| 17  | 1      | 71  | 8190   |
| 19  | 126    | 73  | 504    |
| 23  | 682    | 79  | 8190   |
| 29  | 28     | 83  | 83,886 |
| 31  | 30     | 89  | 2728   |
| 37  | 252    | 97  | 96     |
| 41  | 120    | 101 | 102,300 |
| 43  | 126    | 103 | 510    |

As an exercise, we suggest the following computational project:

**Computational Exploration Project CEP 1** Find the period of the 106-number Ducci game played starting with the binary seed [1, 0, 0, 1, 1, 1, 0, 0, 1, 0, 0, 0, 1, 1, 0, 1, 1, 0, 1, 1, 1, 1, 1, 1, 1, 0, 0, 0, 0, 1, 1, 0, 1, 0, 0, 1, 1, 0, 0, 0, 0, 0, 1, 0, 1, 0, 1, 0, 1, 0, 0, 1, 0, 1, 0, 1, 0, 1, 0, 0, 0, 0, 0, 0, 1, 1, 0, 0, 1, 0, 1, 1, 0, 0, 0, 0, 1, 1, 1, 1, 1, 1, 1, 0, 1, 1, 0, 1, 1, 0, 0, 0, 1, 0, 0, 1, 1, 1, 0, 0, 1] indicating the quadratic residues (1's) and nonresidues (0's) in $F_{107}^*$.

# Chapter 3
# Greatest Prime Factor Sequences

## 3.1 The Prehistory: GPF Sequences, First Contact

Our involvement with "GPF sequences" began at Ohio Northern University in the fall of 2005, out of the necessity of providing one of my students (Lisa Scheckelhoff) with a topic for her senior capstone project. Although I initially thought about something related to the "$3x + 1$" problem (all math students like sequences!), I decided to look for some new problems with a similar flavor but more connected with number theory. So one sunny day, while we were driving our daughter to the ice-skating rink and I was sitting in the back seat of the car alongside our shelty Nana (she was quite an inspiring shelty, all right), I started to scribble sequences of primes with the property that every term is the greatest prime factor of two times the previous term plus one: $x_n = P(2x_{n-1} + 1)$. Just to see what might come out.

I first tried to work with the seed 2, and saw that we quickly get into a cycle $SG$ (from the name of the prolific mathematician Sophie Germain; see below) of length 8:

$$2 \rightarrow \underbrace{[5 \rightarrow 11 \rightarrow 23 \rightarrow 47 \rightarrow 19 \rightarrow 13 \rightarrow 3 \rightarrow 7]}_{SG} \rightarrow 5 \rightarrow \dots$$

Then one would naturally try a seed other than 2 that had not appeared previously. Let's try the smallest such prime. That would be 17, but that would at once bring us back into the same 8-cycle, 7 entering the same cycle $SG$ through the "gate" 7:

$$17 \rightarrow \underbrace{7 \dots}_{SG}.$$

© Springer International Publishing AG 2017
M. Caragiu, *Sequential Experiments with Primes*,
DOI 10.1007/978-3-319-56762-4_3

Next comes 29, which reaches 17 in two iterations, so we will again enter the cycle *SG* through the gate 7:

$$29 \rightarrow 59 \rightarrow 17 \rightarrow \underbrace{7 \ldots}_{SG}.$$

At this point, the smallest seed available for trial is 31, which gets to the cycle element 7 in one step. Then comes 37, which will enter the cycle *SG* in one step, through the gate 5. When we get to 41, we see slightly wilder behavior, getting into the cycle *SG* through the gate 7, after four iterations, reaching a maximum value of 167 along the way:

$$41 \rightarrow 83 \rightarrow 167 \rightarrow 67 \rightarrow \underbrace{5 \ldots}_{SG}.$$

And so on. I communicated the topic to Lisa, and I advised her to get some additional data by trying the calculations by hand first to see whether the pattern held, especially to see whether the cycle *SG* appears to be the only place of convergence. She continued this process by the age-old method of paper and pencil, ultimately covering all the primes less than 1000. The conclusion was for every seed in that range, said GPF iteration is ultimately periodic, leading to the same unique cycle *SG*.

That meant that we had found an interesting experimental math problem to work on, and it was the moment to officially formulate some conjectures, possibly trying to prove something meaningful. The amazing bonus coming from this preliminary verification: in trying to verify similar GPF sequences of the form $x_n = P(ax_{n-1} + b)$, we discovered that they appeared to be always ultimately periodic. We called these "linear" GPF sequences and formulated the following conjecture:

> **The ("linear") GPF Conjecture**
> Every prime sequence defined by a recurrence of the form
> $x_n = P(ax_{n-1} + b)$ is ultimately periodic.

In terms of visuals, one can think of infinite directed graphs $G_{a,b}$ having the prime numbers as vertices and an edge $p \rightarrow q$ whenever $q = P(ax_{n-1} + b)$. The conjecture formulated above can be rephrased as follows: if we follow the flow of directed edges, starting from any vertex in $G_{a,b}$, we eventually get into a limit cycle.

We agreed to call $G_{2,1}$ the "Sophie Germain digraph," since it has an edge for every pair of Sophie Germain primes.

There are plenty of general problems that can be raised. Some can be answered quickly using obvious examples.

Let's consider the question about the existence, in general, of multiple limit cycles. The GPF recurrence $x_n = P(2x_{n-1} + 16)$ produces the limit cycle $[5, 13, 7]$ if we start the recurrence with $x_0 = 29$, and the different limit cycle $[3, 11, 19]$ if we start with $x_0 = 73$.

The next question that can be answered fairly quickly is whether there are values for $a, b$ such that the iteration $x_n = P(ax_{n-1} + b)$ allows exactly one limit cycle. The answer is positive: it is easy to see that the GPF recurrence $x_n = P(x_{n-1} + 1)$ has exactly one limit cycle, [2, 3]. Interestingly, $p \to 2$ is an edge in the associated digraph $G_{1,1}$ if and only if $p$ is a Mersenne prime. As a follow-up to this observation, note that a possible extension of the conjecture about the existence of infinitely many Mersenne primes is the following:

$G_{1,1}$. CONJECTURE : In the digraph, $G_{1,1}$ the in - degree of every vertex is infinite.

Another digraph that can be analyzed in a simple way is $G_{1,2}$, which, for obvious reasons, may be called the "twin-prime digraph." The following simple facts about this digraph are easy to prove:

- There exist a loop $2 \rightleftarrows 2$ (isolated connected component) and a unique cycle (3-cycle) $3 \to 5 \to 7 \to 3$.
- The edge $p \to P(p+2)$ is increasing if and only if $(p, p+2)$ is a twin-prime pair. Moreover, if $(p, p+2)$ is a twin-prime pair, then the edge emerging from $p+2$ is decreasing, with the only exception $p = 3$.
- Starting from any prime $p$, the iterative flow of $G_{1,2}$ leads to the unique aforementioned 3-cycle.

Note the interesting finding that the loop $2 \rightleftarrows 2$ occurs in all GPF digraphs $G_{a,b}$ with the property $b = 2(2^r - a)$. Another example is $G_{5,6}$.

Of course, in addition to the main "linear" GPF conjecture formulated above, difficult questions can be asked about "linear" GPF iterations $x_n = P(ax_{n-1} + b)$:

- In special cases like the "Sophie Germain" iteration $x_n = P(2x_{n-1} + 1)$, we can ask specific questions, such as the following:

  - Is it true that the cycle that constantly appears at the end of every iteration (regardless of the chosen seed) is actually unique?
  - Assuming that the limit cycle for the Sophie Germain iteration is unique, is the distribution of primes that on iteration if chosen as seeds ultimately enter the limit cycle $SG = [5, 11, 23, 47, 19, 13, 3, 7]$ through each of the eight "gates" uniform? That is, if we pick a prime at random as the seed of the iteration $x_n = P(2x_{n-1} + 1)$, is the probability of entering the limit cycle through any one of the eight gates equal to 1/8?
  - Can one provide estimates for the number of iterations needed to enter $SG$.

- In general, for the prime iterations of the form $x_n = P(ax_{n-1} + b)$:

  - Are there always finitely many cycles?
  - Can an upper bound $M$ be provided such that the prime numbers that appear as components of every limit cycle are less than $M$?
  - Can an upper bound $L$ be provided such that the length of every limit cycle can be proved to be less than $L$?

Of course, the "hard questions" most likely need the heavy machinery of analytic number theory, and their solutions are beyond the scope of this book. Most real alternatives for an undergraduate college setting is either providing more computational evidence for the formulated conjectures or to try to prove some special cases.

During the "first contact" with "linear" GPF sequences back in 2005, we used the following MATLAB function to visualize the behavior of the first $n$ terms of a prime sequence $x_n = P(ax_{n-1} + b)$ that starts with a given seed $p$ (for plotting, we use logarithmic plots to smooth the image).

```
function gpfseq = gpfseq(a,b,p,n)
x(1)=p;
for I=2:n;
    x(I)=max(factor(a*x(I-1)+b));
end;
gpfseq=plot(log2(x));
```

The following MATLAB function calculates the number of iterations needed to enter the cycle $SG$:

```
function life = life(m)
J=m;
I=1;
while (J-5)*(J-11)*(J-23)*(J-47)*(J-19)*(J-13)*(J-
3)*(J-7)~=0;
    I=I+1;
    J=max(factor(2*J+1));
end
life=I-1;
```

For example, among the primes less than one million, the one that maximizes the above function is 779,111. If used as a seed in the $x_n = P(2x_{n-1} + 1)$ iteration, we need 19 steps to get into the cycle, pictured below on a logarithmic scale.

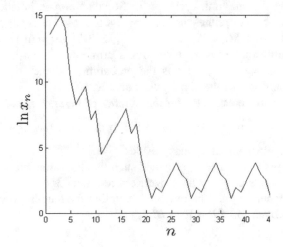

One special elementary case of the "linear" GPG conjecture was proved in the winter of 2005.

**Theorem 3.1** *If $a, b$ are positive integers with $a|b$, then every prime sequence $\{x_n\}_{n \geq 0}$ satisfying the recurrence $x_n = P(ax_{n-1} + b)$ is ultimately periodic.*

A proof of the above theorem begins with a lemma involving the case $a = 1$: every prime sequence $\{x_n\}_{n \geq 0}$ obeying $x_n = P(x_{n-1} + b)$ is ultimately periodic.

In order to do this, we choose to prove a lemma that holds in a more general setting.

**Lemma 3.2** *Let us consider any function $f: \mathbb{N} \to \mathbb{N}$ subject to the following conditions:*

$$f(n) \leq n \quad \text{for all} \quad n, \tag{3.1}$$

$$\lim_{\substack{n \text{ composite} \\ n \to \infty}} (n - f(n)) = \infty. \tag{3.2}$$

*Then we will show that every sequence $\{x_n\}_{n \geq 0}$ that satisfies the recurrence*

$$x_n = f(x_{n-1} + b) \tag{3.3}$$

*is ultimately periodic.*

*Proof of Lemma 3.2* As a consequence of (3.2) and of the fact that arbitrarily large prime gaps exist (as discussed in Section 2.6), there exists a natural number $N > x_0$ such that $f(n) \leq n - b$ for every composite $n \geq N$ and moreover, the integers $N, N+1, \ldots, N+b-1$ are all composite. Under these conditions, we will show that the set $A := \{1, 2, 3, \ldots, N-1\}$ (which contains $x_0$) has the invariance property

$$\forall x[x \in A \Rightarrow f(x+b) \in A].$$

Let $x \in A$. The proof of $f(x+b) \in A$ will proceed by cases:

If $x + b \leq N - 1$, then (3.1) implies $f(x+b) \leq x + b \leq N - 1$, so $f(x+b) \in A$.

If $x + b \geq N$, then $N \leq x + b \leq N + b - 1$ (since $x \leq N - 1$). This makes $x + b$ composite and at least $N$, so that by (3.2), $f(x+b) \leq (x+b) - b = x \leq N - 1$, that is, $f(x+b) \in A$. Lemma 3.2 follows, since the recurrence (3.3) begins from the seed $x_0 \in A$, and since $A$ is invariant under $x \mapsto f(x+b)$, all subsequent terms resulting from the first-order recurrence $x_n = f(x_{n-1} + b)$ are in the finite set $A$, and consequently, $\{x_n\}_{n \geq 0}$ is ultimately periodic.

*Proof of Theorem 3.1* The case $a = 1$ follows from Lemma 3.2, since the greatest prime factor function $P(x)$ satisfies both (3.1) and (3.2); indeed, $P(n) \leq n/2$ for every composite $n$. For the general case $a|b$, say $b = ac$, notice that the GPF recurrence $x_n = P(ax_{n-1} + ac)$ can be rewritten as

$$x_n = \max(P(a), P(x_{n-1} + c)).    \tag{3.4}$$

If the "max" involved in (3.4) produces $P(a)$ twice, the periodicity follows. Otherwise, for all sufficiently large $n$ we have $P(x_{n-1} + c) > P(a)$, so that for $n \geq n_0$, $x_n$ actually satisfies the recurrence $x_n = P(x_{n-1} + c)$, which is by now known to be ultimately periodic. This concludes the proof of Theorem 3.1.

This is, in summary, the work on GPF sequences done in the "first contact" phase. The proof in the more general setting is adapted from a later work of the author (Caragiu 2010). The participating student finished her mathematics senior capstone project in the spring of 2006.

In the period that this undergraduate research session was conducted, Lisa gave a talk on Sophie Germain digraphs at the Nebraska conference of Undergraduate Women in Mathematics, presented an award-winning poster at the 2006 Joint AMS-MAA meetings (undergraduate student poster session), and coauthored a paper (Caragiu and Scheckelhoff 2006) on the topic that included a different elementary proof of periodicity for the prime sequences $\{x_n\}_{n \geq 0}$ satisfying $x_n = P(x_{n-1} + b)$. The proof was based on constructing for every such sequence a certain bound $M$ (specifically, $M = bp + 1$, where $p$ is the least prime that does not divide $b$) with the property that if $x_k > M$, then there exists some $l > k$ with $x_l < x_k$. This results in the (finite) "attractor" set of primes less than $M$ being visited infinitely many times by terms of the sequence, which results in periodicity. Thus, the proof produced another explicit estimate to the effect that at least one element of the limit cycle must be smaller than the aforementioned bound $M$.

Note that as a consequence of Green–Tao's result on the existence of arbitrarily long arithmetic progressions of primes, it turns out that for every $m \geq 1$, there exists a GPF sequence defined by recurrences of the form $x_n = P(x_{n-1} + b)$ that has an increasing segment $x_{i_1} < x_{i_2} < \cdots < x_{i_m}$. The next project was prepared as an additional exercise for students working on MAPLE computations involving the greatest prime factor function.

**Computational Exploration Project CEP 2** Perform a study of the following second-order GPF recurrence:

$$\begin{cases} x_0 = a \\ x_1 = b \\ x_n = P(2x_{n-1} + 1) + P(2P(2x_{n-2} + 1) + 1) \quad \text{for} \quad n \geq 2 \end{cases}.$$

This can be seen as a variation of the Sophie Germain iteration $x_n = P(2x_{n-1} + 1)$.

**Discussion on CEP2** One possibility to get a feeling of these sequences in this project is to "*get some eyes*" with the help of MAPLE while varying the initial conditions $x_0$ **ithprime(rand(1,1000)())** and graphing the first 200 terms (this is flexible and can be varied as needed), we will be able to get a close look at the typical behavior of such a sequence. Here is a sample of what could be seen:

```
> x(0):=ithprime(rand(1..1000)());
x(1):=ithprime(rand(1..1000)()); N:=200: for r from 2 to N do
x(r):= P(2*x(r-1)+1)+P(2*P(2*x(r-2)+1)+1) end do:
X:=[seq(x(r),r=0..N)]; listplot(X);
```

$$x(0) := 1117$$

$$x(1) := 4909$$

$X := [\,1117, 4909, 1114, 802, 1594, 1106, 2922, 400, 156, 492, 216, 512, 58, 96, 196, 174, 612,$
$\quad 240, 42, 22, 12, 16, 22, 28, 30, 74, 190, 150, 60, 40, 26, 60, 118, 102, 94, 90, 186, 384, 852,$
$\quad 50, 108, 60, 18, 60, 16, 34, 46, 78, 164, 54, 128, 330, 764, 146, 324, 646, 448, 886, 244, 242,$
$\quad 206, 72, 46, 90, 188, 40, 62, 12, 16, 22, 28, 30, 74, 190, 150, 60, 40, 26, 60, 118, 102, 94, 90,$
$\quad 186, 384, 852, 50, 108, 60, 18, 60, 16, 34, 46, 78, 164, 54, 128, 330, 764, 146, 324, 646, 448,$
$\quad 886, 244, 242, 206, 72, 46, 90, 188, 40, 62, 12, 16, 22, 28, 30, 74, 190, 150, 60, 40, 26, 60,$
$\quad 118, 102, 94, 90, 186, 384, 852, 50, 108, 60, 18, 60, 16, 34, 46, 78, 164, 54, 128, 330, 764,$
$\quad 146, 324, 646, 448, 886, 244, 242, 206, 72, 46, 90, 188, 40, 62, 12, 16, 22, 28, 30, 74, 190,$
$\quad 150, 60, 40, 26, 60, 118, 102, 94, 90, 186, 384, 852, 50, 108, 60, 18, 60, 16, 34, 46, 78, 164,$
$\quad 54, 128, 330, 764, 146, 324, 646, 448, 886, 244, 242]$

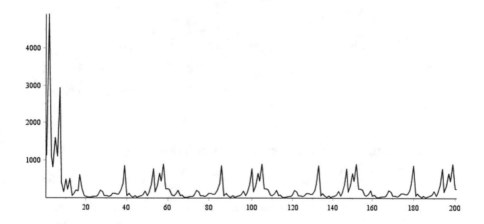

The following preliminary "lab notebook" displays some data about the sequences in CEP2.

| $x_0$ | $x_1$ | Period |
|---|---|---|
| 1117 | 4909 | 47 (most frequent!) |
| 5701 | 733 | 7 |
| 463 | 6029 | 4 |
| 429 | 827 | 3 |
| 3253 | 3691 | 5 |

We discover that a cycle of length 47 occurs most often as output of these "cycle-finding experiments" for sequences in CEP2. Floyd's cycle-finding algorithm with randomized initial conditions (instrumental in acquiring data) would look like this:

```
> X:=(ithprime(rand(1..1000)()),ithprime(rand(1..1000)()));
SLOW:=f(X): FAST:=f(f(X)): k:=1: while SLOW<>FAST do
SLOW:=f(SLOW): FAST:=f(f(FAST)): k:=k+1; end do:
TENTATIVE_START:=k: X:=SLOW: SLOW:=f(X): FAST:=f(f(X)): k:=1:
while SLOW<>FAST do SLOW:=f(SLOW): FAST:=f(f(FAST)): k:=k+1; end
do: PERIOD:=k;
```

$$X := 3253, 3691$$

$$PERIOD := 5$$

**NOTE.** The project CEP2 can obviously be generalized to other higher-order variations of Sophie Germain prime sequences. The sky is the limit! For example, let's try this:

**Computational Exploeation Project CEP3**  Same as in CEP2, only adapt to the following third-order greatest prime factor recurrence:

$$\begin{cases} x_0 = a \\ x_1 = b \\ x_2 = c \\ x_n = P(2x_{n-1}+1) + P(2P(2x_{n-2}+1)+1) + P(2P(2P(2x_{n-3}+1)+1)+1) \\ \qquad\qquad\qquad\qquad\qquad\qquad\qquad\qquad\qquad\qquad\qquad\qquad \text{for} \quad n \geq 3 \end{cases}$$

## 3.2   GPF-Fibonacci: Toward a Generalized GPF Conjecture

The next step in our work came in the period 2009–2010, in joint work with Greg Back, an Ohio Northern junior/senior who majored in computer engineering and applied mathematics in 2010, part of the work being pursued as part of his senior capstone project. The work had two components: the first involved a study of algebraic structures (magmas) constructed with the help of the greatest prime factor function, while the second dealt with generalization of (and acquiring computational evidence for) the ultimate periodicity conjecture to higher-order GPF sequences, including a complete characterization of the limit behavior of a GPF analogue of the Fibonacci sequence, and a fairly extensive computational investigation of the limit cycles of the corresponding GPF analogue of the tribonacci sequence.

In this section we will discuss the first component, which resulted in a paper (Back and Caragiu 2010) that appeared in the *Fibonacci Quarterly* in 2010. It was as much of a "mathematical breakthrough" that a math instructor at a small undergraduate college can hope for, given that our work was referred to in OEIS, while the new sequences were featured in the plenary talk (Sloane 2013) given at the 15th International Conference on Fibonacci Numbers and Their Applications, held in Eger, Hungary, in June 2012.

The generalized Fibonacci sequences $\{G_n\}_{n>0}$ follow the same recurrence relation as the classical Fibonacci sequence, although with different initial conditions:

$$\begin{cases} G_0 = a, \\ G_1 = b, \\ G_n = G_{n-1} + G_{n-2} \ (n \geq 2). \end{cases}$$

For $(a, b) = (0, 1)$, we obtain the traditional $\{F_n\}_{n \geq 0}$. For $(a, b) = (2, 1)$, we obtain the sequence of Lucas numbers $\{L_n\}_{n \geq 0}$. The linearity of the recurrence ensures that for a general set of initial conditions $(a, b)$, the terms of the corresponding sequence $\{G_n\}_{n \geq 0}$ can be written as a linear combination of Fibonacci and Lucas numbers: $G_n = \left(b - \frac{a}{2}\right)F_n + \frac{a}{2}L_n$, so that $G_n$ can consequently be written as a linear combination of the $n$th powers of $\frac{1+\sqrt{5}}{2}$ (the golden section) and its conjugate $\frac{1-\sqrt{5}}{2}$.

Enter the greatest prime factor function! We will consider a class of second-order greatest prime factor sequences $\{x_n\}_{n \geq 0}$ that may be construed as analogues of the generalized Fibonacci sequences. In that class of prime sequences, which we agreed to call *GPF-Fibonacci* sequences, every term is the greatest prime factor of the sum of the two preceding terms:

$$\begin{cases} x_0 = a, \\ x_1 = b, \\ x_n = P(x_{n-1} + x_{n-2})(n \geq 2). \end{cases} \tag{3.5}$$

The parameters $a, b$ (initial conditions) are assumed to be primes. For example, if $a = 2, b = 103$, then the sequence evolves as follows:

$$2, 101, 103, 17, 5, 11, 2, 13, 5, 3, 2, 5, 7, 3, 5, 2, 7, 3, 5, 2. \dots$$

A preliminary verification shows that every such GPF-Fibonacci sequence is ultimately periodic:

- If $a \neq b$, the sequence eventually enters the 4-cycle 7, 3, 5, 2.
- If $a = b$, the sequence is constant: $x_n = a$ for all $n$.

The following elementary observation is instrumental in the proof of the ultimate periodicity:

**Proposition 3.3** *Let $p$ be a prime such that $p + 2$ is composite. Then the initial segment of primes $K_p := \{x | x \text{ prime}, x \leq p\}$ is closed under the operation $(x, y) \mapsto P(x + y)$.*

*Proof* We will begin by observing that $P(x+x) = x$ for every $x \in K_p$. Let $x, y \in K_p$ with $x < y$. If both $x, y$ are odd, then $P(x+y) \in K_p$, since $P(x+y) \leq \frac{x+y}{2} \leq y \leq p$. If $x = 2$, then $P(2+y)$ is clearly less than $p$ if $y < p$, while $P(2+p) < p$ due to the compositeness of $p+2$.

**Corollary** *Let* $p \equiv 1(\text{mod}3)$ *be a prime. Then* $K_p := \{x | x \text{ prime}, x \leq p\}$ *is closed under the operation* $(x, y) \mapsto P(x+y)$.

**Proposition 3.4** *All GPF-Fibonacci sequences are ultimately periodic.*

*Proof of Proposition 3.4* If $\{x_n\}_{n \geq 0}$ is a prime sequence with $x_n = P(x_{n-1} + x_{n-2})$ for $n \geq 2$, let us choose a prime $p \equiv 1(\text{mod}3)$ such that $p \geq \max(x_0, x_1)$. From the above corollary to Proposition 3.3, all terms of $\{x_n\}_{n \geq 0}$ will be confined to the initial prime segment $K_p$. Therefore, every GPF-Fibonacci sequence is ultimately periodic.

Still, this doesn't say anything yet about the possible limit cycles, or whether a unique limit cycle exists. We will begin with a consequence of the fact that $3, 5, 7$ is the only arithmetic progression of primes with difference 2.

**Lemma 3.5** *All the terms of the GPF-Fibonacci sequence defined by* (3.5) *satisfy* $x_n \leq \max(a, b) + 4$. *Moreover, with the single exception of* $\{a, b\} = \{2, 3\}$, *the sharper estimate* $x_n \leq \max(a, b) + 2$ *holds.*

We have seen that if $a = b$ then $\{x_n\}_{n \geq 0}$ is a constant sequence, this trivially gives us a limit cycle of length 1. The following lemma asserts that this cannot happen if $a \neq b$.

**Lemma 3.6** *If* $a \neq b$, *the GPF sequence* (3.5) *cannot enter a limit cycle of length 1.*

*Proof of Lemma 3.6* Let the GPF-Fibonacci sequence $\{x_n\}_{n \geq 0}$ start with $x_0 = a \neq b = x_1$. Let us assume that for some $n$ (which may be taken to be minimal), we have $x_n = x_{n+1} = \cdots = p$. Then $x_{n-1} = p$. Indeed, the recurrence implies $P(x_{n-1} + x_n) = x_{n+1}$, or $P(x_{n-1} + p) = p$. In particular, $p$ divides $x_{n-1} + p$, which means that $p$ divides $x_{n-1}$, and since $x_{n-1}$ is a prime number, it follows that $p = x_{n-1}$, which is a contradiction to the assumed minimality of $n$.

The following is the main result on the prime sequences defined by the GPF-Fibonacci recurrence (3.5).

**Theorem 3.7** *Every GPF-Fibonacci sequence* (3.5) *with* $a \neq b$ *ultimately enters the (unique) limit cycle*

$$[7, 3, 5, 2].$$

*Proof of Theorem 3.7* We have seen in Lemma 3.6 that a sequence (3.5) with $a \neq b$ cannot enter a cycle of length 1. Then we can also rule out a limit cycle of length 2. Indeed, if $[p, q]$ with $p \neq q$ is a limit cycle for $\{x_n\}_{n \geq 0}$, then we quickly get a contradiction, by applying the recurrence (3.5) for the segment $p, q, p, q$, which would imply $P(p+q) = p$ and $P(q+p) = q$, so $p = q$. Thus the limit cycle has length 3 or more. Assume that the limit cycle is

$$L = [p_1, p_2, \ldots, p_k],$$

where $k \geq 3$. We will assume throughout periodic boundary conditions as a notational convention, that is, that $p_{k+1} = p_1, p_{k+2} = p_2, \ldots$. The GPF-Fibonacci recurrence implies that for all i, the following holds:

$$P(p_i + p_{i+1}) = p_{i+2}.$$

Let $M$ be the largest element of the cycle $L$. Say $M = p_m$. Then the cycle elements $p_{m-1}$ and $p_{m-2}$ must be distinct (otherwise, the cycle length would be 1) and both smaller than $M$: if exactly one of $p_{m-1}, p_{m-2}$ is $M$, then $P(p_{m-1} + p_{m-2})$ cannot be equal $M = p_m$. Moreover, one of $p_{m-1}, p_{m-2}$ must be equal to 2. To see this, assume that $p_{m-1}, p_{m-2}$ are odd (and, as already known, distinct and smaller than $M$). Then $M = p_m = P(p_{m-1} + p_{m-2}) \leq \frac{p_{m-1} + p_{m-2}}{2} < M$, a contradiction.

We now distinguish two cases (in each of them, $2 < p < M$).

Case 1: $p_{m-2} = 2$, $p_{m-1} = p \neq 2$, $p_m = M = p + 2$.

In this case, with $p_{m-3} = q \neq 2$, we have $P(q + 2) = p$, and $p - 2 \leq q \leq p + 2$. Since $q = p$ can be easily ruled out, we have either $q = p + 2$ or $q = p - 2$. If $q = p + 2$, then $p = P(q + 2) = P(p + 4)$, a contradiction ($p$ is odd), which shows that $q = p - 2$. Then the arithmetic progression of primes $p - 2, p, p + 2$ necessarily requires $p = 5$ and $M = 7$. Iterating from $p_{m-3} = 3$, we obtain 3, 2, 5, 7, 3, 5, 2, 7, 3, 5, 2, and so on. Thus we see that the actual cycle [7, 3, 5, 2] would begin with $p_m = 7$. That means that although the limit cycle is correctly anticipated, technically, Case 1 never happens, due to the fact that the terms $p_{m-3}, p_{m-2}, p_{m-1}$ would not appear in that particular order within the period.

Case 2 : $p_{m-2} = p \neq 2$, $p_{m-1} = 2$, $p_m = M = p + 2$.

Here we will consider two alternatives.

If $p \leq 11$, we can verify by direct calculation that we eventually enter the cycle 7, 3, 5, 2, with $p = 5$ corresponding to an actual cycle, with $p_{m-2} = 5, p_{m-1} = 2, p_m = 7$ appearing in the precise order.

If $p \geq 13$, since $p, p + 2$ are primes, $p + 4$ is a composite multiple of 3, and thus

$$p_{m+1} = P(p_{m-1} + p_m) = P(p + 4) \leq (p + 4)/3. \tag{3.6}$$

Since $p_m = p + 2$ and $p_{m+1}$ is odd, we therefore have

$$p_{m+2} = P(p_m + p_{m+1}) \leq (p_m + p_{m+1})/2 \leq (4p + 10)/6. \tag{3.7}$$

From Lemma 3.5, the inequalities (3.6) and (3.7), and $p \geq 13$, it follows that

$$p_j \leq \max(p_{m+1}, p_{m+2}) + 4 \leq (4p+34)/6 < p+2 \quad \text{for} \quad j \geq m+1.$$

This means that the maximum cycle value $M = p+2$ will never be attained subsequently, a contradiction that rules out the possibility $p \geq 13$.

Thus the only possibility left for Case 2 is $p \leq 11$, which, as we have seen, corresponds to the generation of the cycle [7, 3, 5, 2] if $p = p_{m-2} = 5$. This completes the proof of Theorem 3.7.

Thus we have found that all Fibonacci-like prime sequences that follow the evolution law $x_n = P(x_{n-1} + x_{n-2})$ and have $x_0 \neq x_1$ are ultimately periodic and enter the nontrivial cycle [7, 3, 5, 2].

An essential ingredient in the proof was the very elementary fact (fitting well with the second-order recurrence) that the greatest prime factor of the sum of two odd primes $p, q$ satisfies $P(p+q) \leq (p+q)/2$. That was instrumental in Propositions 3.3 and 3.4, showing that before anything, the GPF-Fibonacci prime sequences are periodic. The same technique cannot be applied to third-order GPF recurrences such as the class of sequences to be discussed next.

A *GPF-tribonacci* sequence is a prime sequence defined by the third-order recurrence

$$\begin{cases} x_0 = a, \\ x_1 = b, \\ x_2 = c, \\ x_n = P(x_{n-1} + x_{n-2} + x_{n-3}) \quad \text{for} \quad n \geq 3. \end{cases} \tag{3.8}$$

Computational evidence gathered in Back and Caragiu (2010) suggested that these sequences are ultimately periodic. Moreover, the lengths of the limit cycles found by performing a Monte Carlo analysis with various randomly chosen prime values for $a, b, c$ less than 1000 indicated four possible periods (not only in terms of length, but in terms of the nature of the cycles):

- One cycle of length 212, found to occur in 24.05% of cases,
- One cycle of length 100, found to occur in 74.73% of cases,
- Once cycle of length 28, found to occur in 0.76% of cases, and
- One cycle of length 6, found to occur in 0.47% of cases.

Let us try now to improve on this. We will use the power of MAPLE, incorporating Floyd's period-finding algorithm for randomly chosen initial conditions from among the first million primes, for the GPF-tribonacci recurrence (3.6). The third-order recurrence will be implemented by defining the following vector-valued function:

$$f := (x, y, z) \to (y, z, P(x+y+z)).$$

The computer experiment consists of 1000 random trials. As a result of each of them, the resulting period is entered into a period vector "PER," which, in the end, will be sorted so that the number of occurrences of each period will appear. Our first experiment produced 733 occurrences of the period of length 100, 251 occurrences of the period of length 212, 13 occurrences of the period of length 28, and 3 occurrences of the period of length 6.

This allows us to conclude that the proportions of the occurrences of the cycles of length 100, 212, 28, and 6 obtained (73.3, 25.1, 1.3, and 0.3% respectively) was slightly different, but not far, from what was found in Back and Caragiu (2010).

To put it on a slightly more "automatic" regime, we can use the following MAPLE code, which produces the proportions of cycle occurrences of lengths 100, 212, 28, and 6. This computer experiment has two parameters, $T$ (indicating that the random period search is performed within the set of the first $T$ primes), and $M$ (the number of random trials). Each trial will use Floyd's cycle-finding algorithm for the respective random selection of the initial conditions for the GPF-tribonacci recurrence (3.8).

```
> M:=1000: T:=1000000: for r from 1 to M do
X:=(ithprime(rand(1..T)()),ithprime(rand(1..T)()),
ithprime(rand(1..T)())): SLOW:=f(X): FAST:=f(f(X)): k:=1: while
SLOW<>FAST do SLOW:=f(SLOW): FAST:=f(f(FAST)): k:=k+1: end do:
TENTATIVE_START:=k: X:=SLOW: SLOW:=f(X): FAST:=f(f(X)): k:=1:
while SLOW<>FAST do SLOW:=f(SLOW): FAST:=f(f(FAST)): k:=k+1: end
do: PERIOD[r]:=k: end do: PER:=[seq(PERIOD[r],r=1..M)]:
P100:=evalf((Occurrences(100,PER))/M);
P212:=evalf(Occurrences(212,PER)/M);
P28:=evalf(Occurrences(28,PER)/M);
P6:=evalf(Occurrences(6,PER)/M);
is(Occurrences(100,PER)+Occurrences(212,PER)+Occurrences(28,PER)
+Occurrences(6,PER)=M);
```

$$P100 := 0.7420000000$$
$$P212 := 0.2480000000$$
$$P28 := 0.007000000000$$
$$P6 := 0.003000000000$$

*true*

Since we were operating under the hypothesis that 100, 212, 28, and 6 are the only possible limit cycle lengths, we introduced an element of falsification at the end, by checking whether the sum $P100 + P212 + P28 + P6$ equals the parameter $M$ (of course, a "false" output would be sufficient for falsification, but not necessary, since we don't know, for example, whether different limit cycles of length 212 exist). Also, if we feel the need for additional visualization, we can follow up by a histogram with

```
>Histogram(PER, discrete=true);
```

Don't forget to provide the statistics package **with(Statistics)**.

The next table features data from several such GPF-tribonacci experiments, with $T = 1,000,000$ and $M = 1000$. Ideally, in all these prime experiments, increasing the search pool by working within larger search spaces (largest prime sizes, for example) is desirable, because of the general intuitive perception that "higher-energy" experiments may have a chance to produce new "particles" (truths, cycles, …).

| Experiment | P100 | P212 | P28 | P6 |
|---|---|---|---|---|
| 1 | 74.2 | 24.7 | 0.7 | 0.4 |
| 2 | 74.5 | 24.0 | 1.2 | 0.3 |
| 3 | 75.8 | 23.5 | 0.1 | 0.6 |
| 4 | 75.9 | 23.2 | 0.5 | 0.4 |
| 5 | 76.6 | 22.3 | 1.0 | 0.1 |
| 6 | 74.2 | 24.3 | 1.3 | 0.2 |
| 7 | 76.1 | 22.0 | 1.5 | 0.4 |

What about *GPF-tetranacci* sequences?

For the classical tetranacci sequence satisfying the recurrence $x_n = x_{n-1} + x_{n-2} + x_{n-3} + x_{n-4}$ with initial seed 0, 0, 0, 1, see the sequence listed in OEIS as A000078, https://oeis.org/A000078, where one can find some combinatorial interpretations (the $n$th tetranacci number as the number of decompositions of $n - 3$ with no part greater than 4, or the number of bit strings of length $n$ that do not have 1111 as a substring).

With the greatest prime factor function incorporated in the same way as we did for the GPF-tribonacci sequence, a GPF-tetranacci sequence is a prime sequence defined in terms of four parameters (seed $a, b, c, d$) as follows:

$$\begin{cases} x_0 = a, \\ x_1 = b, \\ x_2 = c, \\ x_3 = d, \\ x_n = P(x_{n-1} + x_{n-2} + x_{n-3} + x_{n-4}) \quad \text{for} \quad n \geq 4. \end{cases}$$

Using the same Monte Carlo–style randomized search, the same type of MAPLE experimentation as that used for GPF-tribonacci sequences, we derive computational evidence that all GPF-tetranacci sequences are ultimately periodic. For example, on repeating 10,000 times the process of finding the period of a GPF-tetranacci sequence with randomly chosen initial conditions picked from among the first 1,000,000 primes, the possible periods turn out to be, in decreasing order,

- Period 14 in 68.83% of cases,
- Period 94 in 17.80% of cases,
- Period 3 in 12.85% of cases,
- Period 32 in 0.46% of cases,
- Period 10 in 0.06% of cases.

Repeating seven times the 10,000-trial computer experiment (the larger number of trials was motivated by the small incidence of the case in which the limit cycle has length 10) provided us with the following data:

| Experiment | P14 | P94 | P3 | P32 | P10 |
|---|---|---|---|---|---|
| 1 | 68.46 | 17.95 | 13.11 | 0.45 | 0.03 |
| 2 | 68.14 | 18.26 | 13.07 | 0.46 | 0.07 |
| 3 | 68.12 | 18.22 | 13.07 | 0.51 | 0.08 |
| 4 | 68.32 | 18.87 | 12.31 | 0.43 | 0.07 |
| 5 | 68.06 | 18.55 | 12.82 | 0.45 | 0.12 |
| 6 | 67.48 | 18.92 | 13.15 | 0.44 | 0.01 |
| 7 | 67.88 | 18.74 | 12.75 | 0.59 | 0.04 |

The experiments can continue with GPF analogues of higher-order $n$-step Fibonacci generalizations (Noe et al.) of the classical Fibonacci sequence such as pentanacci numbers (Mendelsohn 1980).

For example, a similar Monte Carlo–type experiment on *GPF-pentanacci* sequences, those satisfying the recurrence $x_n = P(x_{n-1} + x_{n-2} + x_{n-3} + x_{n-4} + x_{n-5})$, reveals limit cycles of lengths 377, 2470, 4935, 9248, and 15,815 (with the last cycle length appearing, at the end of the search, in 94.4% of cases).

**Computational Exploration Project CEP4** Carry out a more detailed analysis on the potentially emergent complexity in terms of possible cycle numbers and cycle lengths for *GPF-k-step-Fibonacci sequences*, that is, prime sequences satisfying the recurrence relation $x_n = P(x_{n-1} + x_{n-2} + x_{n-3} + x_{n-4} + \cdots + x_{n-k})$.

In general, dealing with such higher-order GPF sequences necessitates substantially more computing power. Just to provide a start based on a few concrete examples, we will restrict to a very particular class of such $k$-step GPF-Fibonacci sequences for which the "seed" is an initial segment of the sequence of primes. They are defined by the following "block-sum recurrences" where $p_k$ is the $k$th prime:

$$\begin{cases} x_n = P(x_{n-1} + x_{n-2} + \cdots + x_{n-d}), \\ x_0 = p_1, x_1 = p_2, \ldots, x_{d-1} = p_d. \end{cases}$$

That is, write down the first $d$ primes; thereafter, each subsequent term is the greatest prime factor of the sum of the previous $d$ terms. The idea behind this particular set of sequences was inspired by the Fibonacci sequence, which starts with 0, 1 (the first two nonnegative numbers), with each term the sum of the previous two.

Here are some details on the periods and preperiods for a couple of such sequences:

- The case $d = 2$ leads to a special GPF-Fibonacci sequence (a case settled in Section 3.2 for all possible initial conditions, here 2, 3). This sequence has preperiod 1 and a period of length 4 (specifically, 3, 5, 2, 7).

- The case $d = 3$ is a sequence in the GPF-tribonacci class with period 100 and preperiod 1.
- The case $d = 4$ is a sequence in the GPF-Tetranacci class with preperiod 240 and a period of length 14, specifically, 7, 5, 37, 17, 11, 7, 3, 19, 5, 17, 11, 13, 23, 2.
- The case $d = 5$ is a sequence of period 15815 and preperiod 188347.
- The case $d = 6$ is a sequence of period 2272 and preperiod 2199.
- The case $d = 7$, finalized after about 7 h of Floyd's algorithm running on a Dell Inspiron 17 (with an Intel(R) Core, i7-6500U CPU @ 2.50 GHz and 12 GB RAM) revealed a preperiod of 159384908 and a period of 8266403.

All in all, the following conjecture, initially formulated in Back and Caragiu (2010), seems reasonable to state at this moment:

**The generalized GPF conjecture**
Every prime sequence satisfying a recurrence of the form

$$x_n = P(c_0 + c_1 x_{n-1} + c_2 x_{n-2} + c_3 x_{n-3} + \cdots + c_d x_{n-d})$$

is ultimately periodic.

Returning to first-order recurrences, we can imagine other changes in the recurrence design that may possibly lead to emergent complexity. One possibility is to increase the size of $a, b$ and analyze the corresponding changes in the cycle numbers and sizes. This was tried during the 2011 senior capstone project of Lauren Sutherland, a double major in computer engineering and applied mathematics at Ohio Northern. Multiple limit cycles are the common feature here. For example, the prime sequences satisfying the recurrence relation

$$x_n = P(26{,}390 x_{n-1} + 1103),$$

where the coefficients of the linear argument were chosen, just for fun (or maybe as a lucky charm?), out of Ramanujan's formula for $\pi$, revealed, after a thorough analysis, limit cycles of lengths 18 (occurring in about 10% of cases) and 227 (occurring in about 90% of the cases).

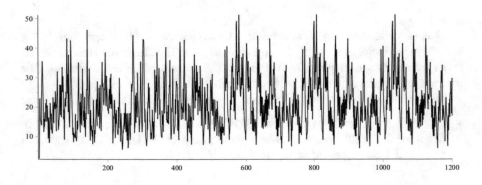

Logarithmic plots of the relevant initial segments for these two sequences are displayed below.

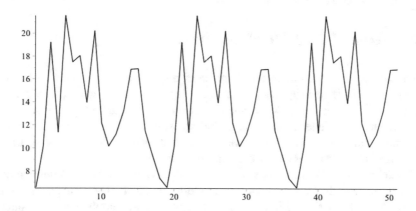

Alternative Explorations

Another alternative to exploring GPF recurrences formulated, in a monolithic way, as $x_n = P(ax_{n-1} + b)$ is described in the following project.

**Computational Exploration Project CEP5**  Use the above technique to take a look at recurrent sequences with sums of first-order greatest prime factor monomials, that is, recurrences of the form $x_n = \sum_{i=1}^{r} c_i P(a_i x_{n-1} + b_i)$. Clearly, the terms of the sequences in this category are not necessarily prime numbers.

Here is one example, a special case, just to whet the appetite: computational evidence suggests that the GPF sequences $x_n = P(2x_{n-1} + 1)$ have one limit cycle $[5, 11, 23, 47, 19, 13, 3, 7]$, and similar evidence can be obtained for GPF sequences $x_n = P(4x_{n-1} + 1)$ that have one limit cycle $[53, 71, 19, 11, 5, 7, 29, 13]$.

However, the first-order recurrences of the form

$$x_n = P(2x_{n-1} + 1) + P(4x_{n-1} + 1)$$

display an increase in complexity, in this class of sequences we find more limit cycles (depending on the seed) of lengths 2, 3, 4, 5, 9, and 60 (with length 60 occurring in the majority of cases explored):

- Length 2: [2920, 11,740]
- Length 3: [604, 2448, 1482]
- Length 4: [2140, 2650, 10,632, 4852]
- Length 5: [76, 78, 470, 960, 280]
- Length 9: [720, 198, 458, 178, 48, 290, 126, 124, 154]
- Length 60: [168, 1010, 496, 728, 1018, 4170, 2822, 1200, 4808, 2300, 3174, 13,604, 120, 278, 610, 2478, 5388, 3908, 8010, 612, 86, 196, 288, 1730, 4230, 25,382, 460, 570, 2444, 8148, 3342, 652, 2638, 1932, 904, 3684, 22,106, 310, 96, 204, 452, 248, 402, 1632, 7182, 28,746, 57,554, 7854, 1036, 1520, 5068, 206, 70, 328, 174, 390, 294, 138, 356, 50]

It would be nice to find measures to quantify the change in complexity on moving from "individual" GPF sequences $x_n = P(a_i x_{n-1} + b_i)$ to sums $x_n = \sum_{i=1}^{r} c_i P(a_i x_{n-1} + b_i)$.

## 3.3  Vector-Valued MGPF Sequences

After looking at "one-dimensional" greatest prime factor sequences defined by recurrences of the form $x_n = P(c_0 + c_1 x_{n-1} + c_2 x_{n-2} + \cdots + c_d x_{n-d})$, in 2011 we concluded that it was time to see whether one could generalize the GPF sequences with which we had been working to multidimensional GPF sequences, or "MGPF" sequences for short. The work was conducted in collaboration with Lauren Sutherland as part of her senior capstone project.

To define MGPF sequences, we use a nonnegative $d \times d$ matrix $A$ with nonzero rows, together with a $d \times 1$ nonnegative column matrix $B$. The terms of an MGPF sequence $\{Q_n\}_{n \geq 0}$ are $d$-dimensional vectors with prime numbers as components:

$$
Q_n = \begin{bmatrix} q_{1,n} \\ q_{2,n} \\ \vdots \\ q_{d,n} \end{bmatrix}.
$$

They satisfy the first-order vector recurrence relation

$$
Q_n = P(AQ_{n-1} + B) \quad \text{for} \quad n \geq 1. \tag{3.9}
$$

The application of the greatest prime factor function $P$ in (3.9) is done componentwise, which means that if $A = [a_{i,j}]_{1 \leq i,j \leq d}$ and $B = [b_i]_{1 \leq i \leq d}$, then (3.9) can be reformulated as the following system of greatest prime factor recurrences:

$$
\begin{cases}
q_{1,n} = P(a_{11}q_{1,n-1} + a_{12}q_{2,n-1} + \cdots + a_{1d}q_{d,n-1} + b_1), \\
q_{2,n} = P(a_{21}q_{1,n-1} + a_{22}q_{2,n-1} + \cdots + a_{2d}q_{d,n-1} + b_2), \\
\cdots \\
q_{d,n} = P(a_{d1}q_{1,n-1} + a_{d2}q_{2,n-1} + \cdots + a_{dd}q_{d,n-1} + b_d).
\end{cases} \tag{3.10}
$$

The seed $Q_0$ is a column vector with $d$ prime components.

The work on MGPF sequences began with a 2011 senior capstone project (Sutherland 2010) at Ohio Northern University. It was followed, in published form, by Caragiu et al. (2011).

Example: The MGPF sequence

$$\begin{bmatrix} q_{1,n} \\ q_{2,n} \end{bmatrix} = P\left( \begin{bmatrix} 2 & 7 \\ 3 & 10 \end{bmatrix} \begin{bmatrix} q_{1,n-1} \\ q_{2,n-1} \end{bmatrix} + \begin{bmatrix} 4 \\ 2 \end{bmatrix} \right), \quad \text{with the initial vector } \begin{bmatrix} q_{1,0} \\ q_{2,0} \end{bmatrix} = \begin{bmatrix} 2 \\ 3 \end{bmatrix}.$$

The periodicity properties of this sequence *for this particular choice of initial condition* can be summarily represented as $Q_{n+7} = Q_n$ for all sufficiently large $n$, and they are distributed to all components of the vectors that appear in the iteration, as detailed below.

- The "upper period," i.e., the period of $\{q_{1,n}\}_{n \geq 0}$, is 7, and the upper limit cycle is $[229, 83, 37, 449, 1801, 21, 757, 1097]$.
- The "lower period," i.e., the period of $\{q_{2,n}\}_{n \geq 0}$, is also 7 (the upper and lower periods are necessarily equal), and the lower limit cycle is $[2341, 277, 53, 643, 2593, 2089, 373]$.

Thus, technically, the limit cycle of the vector-valued MGPF sequence $\{Q_n\}_{n \geq 0}$ subject to the initial condition $Q_0 = \begin{bmatrix} 2 \\ 3 \end{bmatrix}$ is

$$\left[ \begin{bmatrix} 229 \\ 2341 \end{bmatrix}, \begin{bmatrix} 83 \\ 277 \end{bmatrix}, \begin{bmatrix} 37 \\ 53 \end{bmatrix}, \begin{bmatrix} 449 \\ 643 \end{bmatrix}, \begin{bmatrix} 1801 \\ 2593 \end{bmatrix}, \begin{bmatrix} 21,757 \\ 2089 \end{bmatrix}, \begin{bmatrix} 1097 \\ 373 \end{bmatrix} \right].$$

However, the above-mentioned style of Monte Carlo analysis coupled with Floyd's period-finding algorithm reveals as possible periods 6, 7, and 14, with corresponding occurrence percentages 0.4, 92.4, and 7.2%, respectively.

Very interestingly, the period 14 turns out to correspond to two possible non-trivial limit cycles of length 14. The first is

$$\left[ \begin{bmatrix} 2591 \\ 1733 \end{bmatrix}, \begin{bmatrix} 17,317 \\ 5021 \end{bmatrix}, \begin{bmatrix} 821 \\ 283 \end{bmatrix}, \begin{bmatrix} 31 \\ 353 \end{bmatrix}, \begin{bmatrix} 59 \\ 29 \end{bmatrix}, \begin{bmatrix} 13 \\ 67 \end{bmatrix}, \begin{bmatrix} 499 \\ 79 \end{bmatrix}, \right.$$
$$\left. \begin{bmatrix} 311 \\ 109 \end{bmatrix}, \begin{bmatrix} 463 \\ 5 \end{bmatrix}, \begin{bmatrix} 193 \\ 131 \end{bmatrix}, \begin{bmatrix} 1307 \\ 61 \end{bmatrix}, \begin{bmatrix} 29 \\ 1511 \end{bmatrix}, \begin{bmatrix} 10,639 \\ 15,199 \end{bmatrix}, \begin{bmatrix} 5107 \\ 5573 \end{bmatrix} \right].$$

The second limit cycle of length 14 is

$$\left[ \begin{bmatrix} 3 \\ 23 \end{bmatrix}, \begin{bmatrix} 19 \\ 241 \end{bmatrix}, \begin{bmatrix} 19 \\ 823 \end{bmatrix}, \begin{bmatrix} 829 \\ 307 \end{bmatrix}, \begin{bmatrix} 103 \\ 109 \end{bmatrix}, \begin{bmatrix} 139 \\ 467 \end{bmatrix}, \begin{bmatrix} 67 \\ 727 \end{bmatrix}, \right.$$
$$\left. \begin{bmatrix} 5227 \\ 53 \end{bmatrix}, \begin{bmatrix} 17 \\ 523 \end{bmatrix}, \begin{bmatrix} 137 \\ 587 \end{bmatrix}, \begin{bmatrix} 107 \\ 103 \end{bmatrix}, \begin{bmatrix} 313 \\ 41 \end{bmatrix}, \begin{bmatrix} 131 \\ 193 \end{bmatrix}, \begin{bmatrix} 11 \\ 31 \end{bmatrix} \right].$$

Finally, a highly elusive 6-cycle is

$$\left[ \begin{bmatrix} 229 \\ 337 \end{bmatrix}, \begin{bmatrix} 31 \\ 41 \end{bmatrix}, \begin{bmatrix} 353 \\ 101 \end{bmatrix}, \begin{bmatrix} 109 \\ 109 \end{bmatrix}, \begin{bmatrix} 197 \\ 43 \end{bmatrix}, \begin{bmatrix} 233 \\ 31 \end{bmatrix} \right].$$

So, for a single MGPF recurrence relation (in 2D), we have found nontrivial cycles that are distinct (or equivalently, disjoint) and of the same length. Can we do the same for the one-dimensional greatest prime factor sequences? This brings us to the next project.

**Computational Exploration Project CEP6** Find one-dimensional GPF recurrence relations $x_n = P(c_0 + c_1 x_{n-1} + c_2 x_{n-2} + c_3 x_{n-3} + \cdots + c_d x_{n-d})$ that allow for disjoint limit cycles of the same length.

Let us now formulate the multidimensional greatest prime factor conjecture.

**The MGPF conjecture**
Every vector-valued MGPF sequence $Q_n = P(AQ_{n-1} + B)$ is ultimately periodic.

Some particular cases can be proved. The first results from transcribing for MGPFs a particular case of the (one-dimensional) GPF conjecture, which was proved previously in Section 3.2 (Theorem 3.1).

**Proposition 3.8** *The MGPF conjecture is true if* $A = \begin{bmatrix} a_1 & 0 & 0 & 0 \\ 0 & a_2 & 0 & 0 \\ \vdots & \vdots & \ddots & \vdots \\ 0 & 0 & 0 & a_d \end{bmatrix}$ *is a*

*diagonal matrix with positive integers on the diagonal and* $B = \begin{bmatrix} b_1 \\ b_2 \\ \vdots \\ b_d \end{bmatrix}$ *is a non-*

*negative integer vector with* $a_i | b_i$ *for* $i = 1, 2, \ldots, d$.

*Proof of Proposition 3.8* Denote by $Q_n = \begin{bmatrix} q_{1,n} \\ q_{2,n} \\ \vdots \\ q_{d,n} \end{bmatrix}$ the vectors in the MGPF

sequence defined under the given assumptions. If we write $Q_n = P(AQ_{n-1} + B)$ componentwise, we get

$$q_{i,n} = P(a_i q_{i,n-1} + b_i) \quad \text{for} \quad n \geq 1.$$

From the proved special case of the GPF conjecture (feeding $a_i | b_i$ into Theorem 3.1), it follows that the sequences $\{q_{i,n}\}_{n \geq 0}$ are ultimately periodic for $i = 1, 2, \ldots, d$. Say the period of $\{q_{i,n}\}$ is $T_i$.

Then $q_{i,n+T_i} = q_{i,n}$ for all sufficiently large $n$ and every $i = 1, 2, \ldots, d$. Therefore, $Q_{n+T} = Q_n$ for all sufficiently large $n$, where $T = \text{lcm}(T_1, T_2, \ldots, T_d)$. This concludes the proof of Proposition 3.8.

Another elementary case of the MGPF conjecture (Caragiu et al. 2011) is that in which $A$ is a $d \times d$ permutation matrix $A = A_\sigma$ for some permutation $\sigma \in S_d$, that is,

$$a_{1,\sigma(1)} = a_{2,\sigma(2)} = cldots = a_{d,\sigma(d)} = 1 \text{ and } a_{ij} = 0 \text{ in all other positions.}$$

**Proposition 3.9** Let $\sigma \in S_d$. For the permutation matrix $A = A_\sigma$ and a nonnegative $d \times 1$ integer vector B, the prime vector sequence $\{Q_n\}_{n \geq 0}$ satisfying the MGPF recurrence

$$Q_n = P(A_\sigma Q_{n-1} + B) \quad \text{for} \quad n \geq 1$$

is ultimately periodic.

For the proof, we will need first to revisit an elementary result on consecutive composite natural numbers, in the form of the following lemma, which will be instrumental in the proof of Proposition 3.9.

**Lemma 3.10** If $X, X+1, \ldots, X+N-1$ are composite integers, then $X \geq N$.

*Proof of Lemma 3.10* Assume $X < N$. According to the Chebyshev–Bertrand theorem, there exists a prime $p$ such that $X+1 \leq p \leq 2X-1$. This, together with $X < N$, implies $X+1 \leq p \leq x+N$, in contradiction to the compositeness assumption.

*Proof of Proposition 3.9* The greatest prime factor recurrence formula $Q_n = P(A_\sigma Q_{n-1} + B)$ can be written, componentwise, as

$$q_{i,n} = P(q_{\sigma(i),n-1} + b_i) \quad \text{for} \quad n \geq 1. \tag{3.11}$$

Let $M := \max(q_{1,0}, q_{2,0}, \ldots, q_{d,0})$ and $L := \max(b_1, b_2, \ldots, b_d)$.

If $L = 0$, then (3.11) becomes $q_{i,n} = P(q_{\sigma(i),n-1}) = q_{\sigma(i),n-1}$ for $n \geq 1$, so that every subsequent $Q_n$ can be found by permuting the $d$ components $Q_{n-1}$ through the action of $\sigma \in S_d$. Then $\{Q_n\}_{n \geq 0}$ is periodic, with period at most of the order of $\sigma \in S_d$.

If $L > 0$, let $X, X+1, \ldots, X+L-1$ be a sequence of composite integers such that

$$X > M = \max(q_{1,0}, q_{2,0}, \ldots, q_{d,0}).$$

Note that Lemma 3.10 ensures that $X \geq L$. We will prove by induction on $n$ that

$$q_{i,n} \leq X \quad \text{for} \quad i = 1, 2, \ldots, d.$$

If   $n = 0$,   then   $q_{i,0} \leq X$   from   the   choice   of   $X$.   Assuming $q_{i,n-1} \leq X$   for   $i = 1, 2, \ldots, d$, we will show that

$$q_{i,n} = P(q_{\sigma(i),n-1} + b_i) < X \tag{3.12}$$

for all $i = 1, 2, \ldots, d$.

Indeed (3.12) trivially follows if $q_{\sigma(i),n-1} + b_i < X$. Let us assume, for some $i = 1, 2, \ldots, d$, that

$$q_{\sigma(i),n-1} + b_i \geq X.$$

But $q_{\sigma(i),n-1} < X$ by the definition of $X$ and $b_i \leq L$ by the definition of $L$. Then

$$X \leq q_{\sigma(i),n-1} + b_i < X + L. \tag{3.13}$$

Clearly by the choice of $X$ and $L$, $q_{\sigma(i),n-1} + b_i$ is composite. Consequently,

$$q_{i,n} = P(q_{\sigma(i),n-1} + b_i) \leq \frac{q_{\sigma(i),n-1} + b_i}{2} < \frac{X+L}{2} \leq \frac{X+X}{2} = X \text{ for all } i = 1, 2, \ldots, d.$$

The ultimate periodicity announced in Proposition 3.9 follows now, since the set

$$U = \left\{ \begin{bmatrix} x_1 \\ x_2 \\ \vdots \\ x_d \end{bmatrix} \middle| x_i \in \mathbb{N} \text{ and } x_i < X \text{ for } 1 \leq i \leq d \right\} \text{ is finite, and } Q_n \in U \text{ for all } n \geq 0.$$

Moving slightly beyond permutation matrices, we will consider another category of sparse matrices, namely 0/1 matrices with at most two 1's per row. The next result (Caragiu et al. 2011) conveys another special case in which the MGPF conjecture is true.

**Proposition 3.11** *The MGPF conjecture is valid for $B = 0$ and $d \times d$ matrices $A$ with 0/1 entries, nonzero rows, and at most two 1's in each row.*

*Proof of Proposition 3.11* Under the stated assumptions, the associated MGPF iteration sends a prime vector $X = \begin{bmatrix} q_1 \\ q_2 \\ \vdots \\ q_d \end{bmatrix}$ into the vector $P(AX) = \begin{bmatrix} r_1 \\ r_2 \\ \vdots \\ r_d \end{bmatrix}$, where every component $r_i$ is a sum of at most two $q_j$'s. Let us select a prime $p \equiv 1 \pmod{3}$ such that all the components of the seed $X_0$ are less than $p$. Recall (Proposition 3.3) that the initial prime segment $K_p = \{x | x \text{ prime}, x \leq p\}$ is closed under the operation consisting of the greatest prime factor of the sum $(x, y) \mapsto P(x + y)$. The closure property implies that for sequences of this particular kind, all components of the subsequently produced prime vectors $X_1, X_2, \ldots$ are in $K_p$. Since all terms of the vector-valued MGPF sequence are confined to the finite set $K_p^d$, the ultimate periodicity follows, which concludes the proof of Proposition 3.11.

**Corollary** *In particular, the MGPF conjecture is valid for $B = 0$ and $d \times d$ circulant matrices $A$ of the form*

$$
A = \begin{bmatrix} 1 & 1 & \cdots & 0 & 0 & 0 \\ 0 & 1 & 1 & 0 & \cdots & 0 \\ \cdots & \cdots & \cdots & \cdots & \cdots & \cdots \\ 0 & 0 & 0 & \cdots & 1 & 1 \end{bmatrix}.
$$

*This can be construed as a variation of the d-number Ducci game, in which the Ducci iteration*

$$
(x_1, x_2, x_3, \ldots, x_{d-1}, x_d) \mapsto (|x_1 - x_2|, |x_2 - x_3|, |x_3 - x_4|, \ldots, |x_{d-1} - x_d|, |x_d - x_1|)
$$

*is replaced by* $(x_1, x_2, x_3, \ldots, x_{d-1}, x_d) \mapsto (|x_1 - x_2|, |x_2 - x_3|, |x_3 - x_4|, \ldots,$ $|x_{d-1} - x_d|, |x_d - x_1|)$. *We will see that in this special case, much more can be said about the limit cycles.*

Note that for $d > 1$, the $d$-dimensional MGPF conjecture implies the GPF conjecture: $\text{MGPF}_d \Rightarrow \text{GPF}$. Indeed, it is easy to see this: every GPF sequence $(q_n)_{n \geq 0}$ satisfies the $d$-dimensional MGPF recurrence

$$
\begin{bmatrix} q_n \\ q_{n-1} \\ \vdots \\ q_{n-d+1} \end{bmatrix} = P \left( \begin{bmatrix} a & 0 & \cdots & 00 \\ 1 & 0 & \cdots & 00 \\ 0 & 1 & \cdots & 00 \\ 0 & 0 & \cdots & 10 \end{bmatrix} \begin{bmatrix} q_{n-1} \\ q_{n-2} \\ \vdots \\ q_{n-d} \end{bmatrix} + \begin{bmatrix} b \\ 0 \\ \vdots \\ 0 \end{bmatrix} \right).
$$

It is not clear, however, whether the converse is true. There is a great deal of "mixing" in the general multidimensional greatest prime factor iterations.

We discussed vector-valued greatest prime factor sequences. But needless to say, those vectors could be matrices. We can write the MGPF construction in just about the same way for matrix-valued greatest prime factor sequences (lucky for us, it's the same acronym). That is, we are talking about sequences of matrices $\{M_n\}_{n \geq 0}$ with prime numbers as entries satisfying a recurrence of the form $M_n = P(AM_{n-1} + B)$, where this time, $A, B$ are both $d \times d$ matrices, and $P$ applies componentwise to any matrix with natural number components.

## MGPFs and Graphs

One way to visualize vector-valued greatest prime factor sequences is with graphs.

This graph recurrence will involve simple undirected graphs with vertices labeled by primes. A step in the graphical iteration will update the label at each vertex $v$ by replacing it with the greatest prime factor of the sum (or the weighted sum) of the labels of vertices in a certain graph neighborhood of $v$.

Let's take a simple example, based on the following graph:

We will look at it as a 3-by-3 grid with cells naturally representing the cardinal/intercardinal directions, together with the middle, denoted by M. Let's define the "neighborhood" of a cell C to consist of the cell itself together with the cells sharing a common edge with C. For example, the neighborhood of W will consist of W, NW, SW, and M, while the neighborhood of SE consists of SE, E, and S.

| NW | N | NE |
|----|---|----|
| W  | M | E  |
| SW | S | SE |

The process of iteration will start with a "seed"; that is, we will choose a prime number for each cell $x_{m,0}, x_{n,0}, x_{ne,0}, x_{e,0}, x_{se,0}, x_{s,0}, x_{sw,0}, x_{w,0}, x_{nw,0}$, and then proceed with sequential updates of the form

- $x_n \mapsto P(x_m + x_n + x_{nw} + x_{ne})$
- $x_{ne} \mapsto P(x_{ne} + x_n + x_e)$
- $x_e \mapsto P(x_m + x_{ne} + x_e + x_{se})$
- $x_{se} \mapsto P(x_{se} + x_s + x_e)$
- $x_s \mapsto P(x_m + x_s + x_{se} + x_{sw})$
- $x_{sw} \mapsto P(x_{sw} + x_s + x_w)$
- $x_w \mapsto P(x_m + x_w + x_{nw} + x_{sw})$
- $x_{nw} \mapsto P(x_{nw} + x_n + x_w)$
- $x_m \mapsto P(x_m + x_n + x_e + x_s + x_w)$

Again, using the search for the MGPF iteration periods using Floyd's algorithm together with a Monte Carlo–style walk through a randomized selection of seeds, we detect the following possible periods for the graphical iteration defined above: 1 (let's imagine why such a cycle length might occur), 4, 30, 104, and 1541.

Before we break for an algebraic intermezzo, here is another project.

**Computational Exploration Project CEP7** After exploring multidimensional GPF recurrences, let's take a step toward "nonlinear" ones (Caragiu 2010). This is in general much harder, so at the beginning, let's stay close to the "linear" greatest prime factor sequences, subject to a recurrence of the form $x_n = P(ax_{n-1} + b)$.

To this end, consider a polynomial $f(x) \in \mathbb{Z}[x]$ of degree $k$ that splits as a product of linear factors with natural number coefficients, $f(x) = (a_1 x + b_1)(a_2 x + b_2)\ldots(a_k x + b_k)$ with $c, a_i, b_i \in \mathbb{N}$. Explore and provide computational evidence for the conjecture that every prime sequence that satisfies the nonlinear (albeit misleadingly nonlinear, due to the assumed factorization) $x_n = P(f(x_{n-1}))$ for $n \geq 1$ is ultimately periodic, and acquire data on its limit cycle structure.

Note that such a "split-nonlinear" recurrence relation $x_n = P(f(x_{n-1}))$ can be rewritten as

$$x_n = \max(P(a_1 x_{n-1} + b_1), P(a_2 x_{n-1} + b_2), \ldots, P(a_k x_{n-1} + b_k)).$$

It is not clear whether the ultimate periodicity of such split-nonlinear GPF sequences follows conditionally if one assumes the ultimate periodicity of "linear" GPF recurrences of the form $x_n = P(ax_{n-1} + b)$.

Here are a few examples:

- Computational evidence points to a single 8-cycle for the GPF recurrence $x_n = P(2x_{n-1} + 1)$ and a single 8-cycle for $x_n = P(4x_{n-1} + 1)$, while for the recurrence $x_n = P((2x_{n-1} + 1)(4x_{n-1} + 1)) = P(8x_{n-1}^2 + 6x_{n-1} + 1)$, it reveals two limit cycles, of lengths 13 (in about 54.7% of cases) and 14 (in about 45.3% of cases).
- In addition, computational evidence points to two cycles of lengths 11 and 19 for the recurrence $x_n = P(6x_{n-1} + 1)$. Combining the three above results shows that the GPF sequence $x_n = P((2x_{n-1} + 1)(4x_{n-1} + 1)(6x_{n-1} + 1)) = P(48x_{n-1}^3 + 44x_{n-1}^2 + 12x_{n-1} + 1)$ displays, as a result of the randomized application (in terms of initial conditions) of Floyd's algorithm, limit cycles of lengths 9 (in 9.6% of cases) and 106 (in 91.4% of cases).
- One may ask whether the GPF sequences in this split-nonlinear category have multiple periods. This is probably not true: the fifth-degree GPF recurrence $x_n = P((2x_{n-1} + 1)(4x_{n-1} + 1)(8x_{n-1} + 1)(16x_{n-1} + 1)(32x_{n-1} + 1))$ appears to have only one limit cycle, of length 740, with maximum cycle element 18865553077448503.

We conclude this section with a logarithmic plot of an initial segment of the GPF sequence $x_n = P((2x_{n-1} + 1)(4x_{n-1} + 1)(8x_{n-1} + 1)(16x_{n-1} + 1)(32x_{n-1} + 1))$ with $x_0 = 8101$.

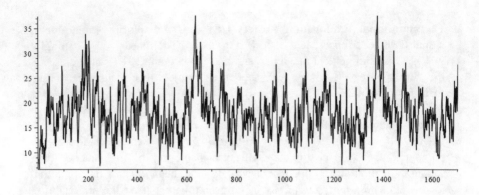

## 3.4    The Ubiquitous 2, 3, 5, 7 and an Interesting
         Magma Structure

In what follows, we would like step aside for a while from the topic of prime
sequences and proceed with an intermezzo on a class of nonassociative algebraic
structures (magmas) on the set of the primes.

The reason for our intermezzo is actually connected to the topic of greatest prime
factor sequences. In the course of the proof of the ultimate periodicity of the
second-order GPF-Fibonacci recurrence $x_n = P(x_{n-1} + x_{n-2})$, Proposition 3.4,
which asserts the closure under the binary operation $(x,y) \mapsto P(x+y)$ of the set of
primes $K_p = \{x | x \text{ prime}, x \leq p\}$, where $p$ is a prime such that $p \equiv 1 \pmod 3$, was
instrumental in the proof of the validity of the final result.

So the following idea came along quite naturally. Let $\Pi$ be set of primes. Let us
define the following binary operation on $\Pi$:

$$x * y = P(x+y)$$

for all $x, y \in \Pi$.

The structure $(\Pi, *)$ is an infinite algebraic structure that is commutative (obviously $x * y = y * x$) but nonassociative. To see this, note that

$$2 * (2 * 3) = 2 * 5 = 7 \neq 5 = 2 * 3 = (2 * 2) * 3.$$

Also, $(\Pi, *)$ has no identity of any kind: no prime $p$ satisfies $p * x = x$, that is,
$P(p + x) = x$ for every $x \in \Pi$ (actually, $p * x = x$ holds only for $x = p$). Aside from
commutativity and idempotence, that is, $x * x = x$ for all $x \in \Pi$, the algebraic
structure $(\Pi, *)$ has no particularly notable algebraic properties. A generic algebraic
structure defined by a set with a binary operation, even an operation that might

seem more or less "random," is called a *magma* or a *groupoid*. Of course, sets with highly structured binary operations, such as groups, are also magmas.

In the language of structures, Proposition 3.3 can be restated as follows:

If $p \equiv 1 \pmod 3$, then $K_p$ is a submagma of $(\Pi, *)$.

To put this to good use, let us note one interesting consequence, that the magma $(\Pi, *)$ is not finitely generated. Indeed, assume that $(\Pi, *)$ is generated by a finite set of primes $X$. Selecting a large enough prime $p \equiv 1 \pmod 3$ such that $X \subseteq K_p$, it would follow that $\langle X \rangle \subseteq K_p$, so $\langle X \rangle$ cannot equal $\Pi$.

The smallest $K_p$-type submagma of the structure $(\Pi, *)$ is $K_7 = \{2, 3, 5, 7\}$, which will play an important role in what follows. For the moment, here is its multiplication table:

| * | 2 | 3 | 5 | 7 |
|---|---|---|---|---|
| 2 | 2 | 5 | 7 | 3 |
| 3 | 5 | 3 | 2 | 5 |
| 5 | 7 | 2 | 5 | 3 |
| 7 | 3 | 5 | 3 | 7 |

Incidentally, this is an interesting example of an algebraic structure that can supplement an elementary abstract algebra class.

More generally, we can define a class of algebraic structures $(\Pi, f_{a,b})$ on the set of primes depending on two parameters $a, b \in \mathbb{N}$ by setting

$$f_{a,b}(x, y) := P(ax + by) \quad \text{for} \quad x, y \in \Pi.$$

Note that $(\Pi, f_{1,1})$ is none other than $(\Pi, *)$. The magmas $(\Pi, f_{a,b})$ are always nonassociative and, except for the case $a = b$, noncommutative. This results from the following proposition (Caragiu and Back 2009).

**Proposition 3.12** *There exists a prime $p \geq a + b$ such that*

$$f_{a,b}(f_{a,b}(p, p), 2) \neq f_{a,b}(p, f_{a,b}(p, 2)).$$

*Moreover, if $f_{a,b}(p, 2) = f_{a,b}(2, p)$ for every $p \in \Pi$, then $a = b$.*

*Proof of Proposition 3.12* Let us assume that

$$f_{a,b}(f_{a,b}(p, p), 2) = f_{a,b}(p, f_{a,b}(p, 2)) \quad \text{for all} \quad p \geq a + b. \tag{3.14}$$

Note that $f_{a,b}(p, p) = P(ap + bp) = P(p(a + b)) = p$ for $p \geq a + b$, so (3.14) can be rewritten as $f_{a,b}(p, 2) = f_{a,b}(p, f_{a,b}(p, 2))$. Thus,

$$P(ap + 2b) = P(ap + bP(ap + 2b)) \tag{3.15}$$

holds for every $p \geq a + b$. Let $q := P(ap + 2b)$. Then clearly, $q$ (the greatest prime divisor of $ap + 2b$) satisfies $q = P(ap + bq)$ according to (3.15), so $q|ap$, and since $q|(ap + 2b)$, it follows that $q|2b$, or

$$P(ap + 2b)|2b \tag{3.16}$$

for all sufficiently large primes $p$. But this cannot happen. To see why, select a prime $r$ such that $r > \max(a, 2b)$. By Dirichlet's theorem for primes in arithmetic progressions, there are infinitely many primes $p$ satisfying the congruence

$$ap \equiv -2b \pmod{r} \tag{3.17}$$

For every prime $p$ satisfying (3.17), we have $P(ap + 2b) \geq r > 2b$. But this is in contradiction to (3.16). This concludes the first part of the proof.

For the remaining part, let us assume that $a \neq b$ and $f_{a,b}(p, 2) = f_{a,b}(2, p)$ for every $p \in \Pi$. In other words,

$$P(ap + 2b) = P(2a + bp)$$

for every prime $p$. Let $s := P(ap + 2b) = P(2a + bp)$. Since $s|(ap + 2b)$ and $s|(2a + bp)$, it follows that $s$ will divide their difference, or

$$s|(a - b)(p - 2).$$

For the remaining part, let us assume that $a \neq b$ yet $f_{a,b}(p, 2) = f_{a,b}(2, p)$ for every $p \in \Pi$. In other words,

$$P(ap + 2b) = P(2a + bp)$$

for every prime $p$. Let

$$s := P(ap + 2b) = P(2a + bp). \tag{3.18}$$

Since $s|(ap + 2b)$ and $s|(2a + bp)$, it follows that $s$ will divide their difference, or

$$s|(a - b)(p - 2). \tag{3.19}$$

Adapting the argument provided in the first part, let us choose a prime $r > 2a + 2b$. Then for every prime $p$ satisfying the congruence (solvable for primes, according to Dirichlet's theorem) $ap \equiv -2b \pmod{r}$, we will have

$$s = P(ap + 2b) \geq r > 2a + 2b > |a - b|. \tag{3.20}$$

Therefore, from (3.19) and (3.20), for such a choice of $p$, the prime $s$ defined by (3.18) will divide $p - 2$, and thus $s$ will divide the linear combination $-a(p - 2) + (ap + 2b) = 2(a + b)$, which is in contradiction to (3.20), which has

$s > 2a + 2b$. Thus if $a \neq b$, it is impossible to have $f_{a,b}(p,2) = f_{a,b}(2,p)$ for every $p \in \Pi$, so that $(\Pi, f_{a,b})$ is commutative if and only if $a = b$.

This concludes the proof of Proposition 3.12.

At this moment, seeing that we sensed we might get into a zone of potentially heavy nonassociative algebra content, since none of us had any significant experience in the mainstream theory, we looked at several classic textbooks on nonassociative algebra, but going over the material seemed too hard and time-consuming for us. Consequently, we decided to play with the nonassociative content emerging around these nonassociative magmas with an ab initio state of mind, self-contained and elementary as it came to us, with the hope that everybody would gain from this ad hoc experiential learning context.

Let's take a closer look at the case $a = b = 1$, that is, at the magma $(\Pi, *)$, especially at its submagma structure.

Clearly, due to the idempotence of the operation $*$, every 1-element set $\{x\}$ with $x \in \Pi$ constitutes a submagma. Let's agree to call these one-element submagmas trivial and explore the structure of the nontrivial submagmas of $(\Pi, *)$. We already introduced one of them, $K_7 = \{2,3,5,7\}$. The next result (Caragiu and Back 2009) will spell out a reason why $\{2,3,5,7\}$ is indeed a special submagma.

**Theorem 3.13** *The submagma $K_7 = \{2,3,5,7\}$ is included in every nontrivial submagma $Y$ of $(\Pi, *)$.*

*Proof of Theorem 3.13* Let $Y$ be a nontrivial submagma of $(\Pi, *)$. The proof of $\{2,3,5,7\} \subseteq Y$ will be conducted in four steps.

*Firstly,* we show that $2 \in Y$. Indeed, if $2 \notin Y$ and $x, y \in Y$ with $x < y$ are the least two (necessarily odd) elements of $Y$, then the element $x * y$ of $Y$ (recall that $Y$ is closed under the operation $*$) satisfies

$$x * y = P(x+y) \leq \frac{x+y}{2} < y.$$

This implies $P(x+y) = x$ with $x, y$ prime, which is possible only if $x = y$, contradicting the choice of $x, y$. This concludes the first step. Now we know that $2 \in Y$.

*Secondly,* we show that $Y$ must contain at least two more elements besides 2. Indeed, if $Y = \{2,x\}$ with $x$ an odd prime, then $2 + x$ is odd, so we must necessarily have $2 * x = x$, but that would imply $x|2$, a contradiction.

*The third step* will reveal the identity of the minimal two *odd* elements of $Y$ (now that we know of their existence), by showing that if $\{2,x,y\} \subseteq Y$ with $x, y$ odd and $x < y$, then $x = 3$ and $y = 5$. Indeed, if $x, y$ are odd elements of $Y$, then $x * y = P(x+y)$ is distinct from both $x$ and $y$, and as before, it satisfies $x * y \leq \frac{x+y}{2} < y$. The only possibility left is $x * y = 2$, that is,

$$x * y = 2^k \tag{3.21}$$

for some $k \geq 1$. But since $2 * x = P(x+2)$ cannot possibly be in the set $\{2,x\}$ and is at most $x + 2 \leq y$, it necessarily follows that $2 * x = y$ and

$$x + 2 = y. \tag{3.22}$$

We will now show that if $x < y$ are odd primes satisfying (3.21) and (3.22), then $x = 3$ and $y = 5$. Indeed, from (3.21) and (3.22) it follows that

$$x = 2^{k-1} - 1 \quad \text{and} \quad y = 2^{k-1} + 1$$

for some $k \geq 3$. The above relation makes $x$ a Mersenne prime (implying that $k - 1$ is prime) and $y$ a Fermat prime (implying that $k - 1$ is a power of 2); see Section 2.5 in Hardy and Wright (1979). Since the only possibility for $k - 1$ being both a prime and a power of 2 is $k - 1 = 2$, it follows that $k = 3$, and thus $x = 3$, $y = 5$, whence $\{2, 3, 5\} \subseteq Y$. Since $Y$ is closed under the operation $*$, it follows that $2 * 5 = 7 \in Y$. Therefore, $K_7 = \{2, 3, 5, 7\} \subseteq Y$, which concludes the proof of the fact that $K_7$ is included in every nontrivial submagma of $(\Pi, *)$.

We have seen that if $p$ is a prime with $p + 2$ composite, then $K_p = \{x | x \leq p \text{ and } x \text{ prime}\}(\Pi, *)$. A result in the opposite direction is the following.

**Proposition 3.14** *Let $Y$ be a finite nontrivial submagma of $(\Pi, *)$, and let $p$ be its maximal element. Then the element $p + 2$ is composite.*

*Proof of Proposition 3.14* Since (from Theorem 3.13) we have $2 \in Y$, it follows that $p * 2 = P(p + 2) \in Y$. Then $p + 2$ must be composite, since, otherwise the assumed maximality of $p$ as an element of $Y$ would be contradicted.

The following question naturally arises at this moment:

*Is every submagma of $(\Pi, *)$ a $K_p$ ?*

A brief inspection shows that the answer is negative.

Indeed, the set $\{2, 3, 5, 7, 11, 13, 17, 19, 31\}$ is a submagma of $(\Pi, *)$ not of the form $K_p$. In fact, this 9-element submagma is the $*$-closure of the 2-element set $\{2, 31\}$ (a good exercise for both faculty and student would be to try your hand at such statements, in a good old "paper and pencil" mode).

Where should we go from here? Inspired by the above counterexample, we define a new class of nontrivial submagmas of $(\Pi, *)$: for an odd prime $p$, consider the closure

$$C_p := \overline{\{2, p\}}$$

Clearly, $C_p$ is nontrivial and thus contains $\{2, 3, 5, 7\}$ for $p \geq 3$, in which case

$$C_p := \overline{\{2, p\}} = \overline{\{2, 3, 5, 7, p\}}$$

Note that every nontrivial submagma $Y$ of $(\Pi, *)$ can be written as a union of special submagmas of the form $C_p$, in the following way: $Y = \bigcup_{\substack{p \in Y \\ p \text{ odd}}} C_p$.

Once we have submagmas of the form $K_p$ $(p \geq 7)$ and submagmas of the form $C_p$, we will have to look for some relationships among them.

**Proposition 3.15** *If $p \geq 5$ is a prime, then*

1. $C_p \subseteq K_p$ *if $p+2$ is composite,*
2. $C_{p+2} \subseteq C_p \subseteq K_{p+2}$ *if $p+2$ is prime.*

*Proof of Proposition 3.15* Let $p \geq 5$ be a prime. If $p+2$ is composite, then $p \geq 7$ and $K_p$ is a submagma with $\{2,p\} \subset K_p$. Therefore, the relation $C_p = \overline{\{2,p\}} \subseteq K_p$ holds, which proves the first part.

If $p+2$ is prime, then $p+2 = p * 2 \in C_p$ and $p+4$ is composite, which makes the existence of $K_{p+2}$ possible. Since $\{2,p+2\} \subset C_p$, it follows that $C_{p+2} = \overline{\{2,p+2\}} \subseteq C_p$. On the other hand, $K_{p+2}$ is a submagma that contains 2 and $p$, which results in $C_p = \overline{\{2,p\}} \subseteq K_{p+2}$, which concludes the proof of the second part.

Let's consider the status of twin prime pairs. The following result has some relevance in this context.

**Proposition 3.16** *Let $(p,p+2)$ with $p > 5$ be a twin prime pair. Then $p \notin C_{p+2}$, and consequently, the inclusion $C_{p+2} \subseteq C_p$ proved in Proposition 3.15 is strict.*

*Solution to Proposition 3.16* Since $(p,p+2)$ is a twin prime pair and since $p > 5$, both $p-2$ and $p+2$ are composite. Assuming $p \in C_{p+2} = \overline{\{2,p+2\}}$, $p$ must be of the form

$$p = m * n = P(m+n)$$

for some distinct primes $m, n \in C_{p+2}$ with $p \notin \{m,n\}$ (here we also used the idempotence of the magma operation). Without loss of generality, assume $m < n$.

We will prove that $m \neq 2$ by showing that if $m = 2$, then $P(m+n) = P(2+n) < p$. Indeed, this is obvious if $n < p$ (because then $n < p-2$ due to the compositeness of $p-2$) and also if $n = p+2$(because then $n+2 = p+4$ is composite and $P(2+n) = P(p+4) < (p+4)/3 < p$).

Therefore we may assume that the two primes $m < n$ are be odd with $m < p-2$ and $n < p+2$. Then $m+n$ is even and $m+n < 2p$, which implies $m * n = P(m+n) \leq (m+n)/2 < p$, a contradiction. This concludes the proof of Proposition 3.16.

After this group of elementary results, let's shift gears a little bit toward a more experimental approach. One possible question is the following.

*If $p$ is a prime, describe the proportion of primes up to $p$ that can be generated from 2 and $p$ using the magma operation "$*$" repeatedly.*

Our choice (Caragiu and Back 2009) was to explore the "density function"

$$D(p) := \frac{|C_p \cap K_p|}{|K_p|} = \frac{|C_p \cap K_p|}{\pi(p)}. \tag{3.23}$$

Although not of interest to the asymptotic estimates, small values of $p$ of the density function are $D(3) = D(5) = D(7) = 1$, and by an ad hoc convention, we say that $D(2) := 1$. If $p > 7$, we distinguish two cases:

A.  If $p+2$ is composite, then $C_p \subseteq K_p$, and therefore

$$D(p) = \frac{|C_p|}{|K_p|} = \frac{|C_p|}{\pi(p)}.$$

B.  If $p+2$ is prime, then $C_p - K_p = \{p+2\}$, and therefore we have

$$D(p) = \frac{|C_p \cap K_p|}{|K_p|} = \frac{|C_p| - 1}{\pi(p)}.$$

A rephrasing of the twin-prime-related result Proposition 3.16 in terms of the submagma densities just defined is the following.

**Proposition 3.17**  *If $(p, p+2)$ is a twin prime pair with $p \geq 7$, then*

$$D(p+2) < D(p).$$

*Proof of Proposition 3.17*  For a twin prime pair $(p, p+2)$ we have seen that $C_{p+2} \subset C_p$ (a strict inclusion, from Proposition 3.16), so that

$$D(p) = \frac{|C_p| - 1}{\pi(p)} \geq \frac{|C_{p+2}|}{\pi(p)} > \frac{|C_{p+2}|}{\pi(p) + 1} = \frac{|C_{p+2}|}{\pi(p+2)} = D(p+2),$$

where in the last equality we used the fact that $p+4$ is composite.

The MATLAB graph below (Caragiu and Back 2009) displays the values of the density function $D(p_n)$, where $p_n$ is the $n$th prime, for $5 \leq n \leq 250$.

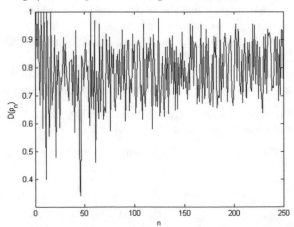

**Computational Exploration Project CEP8** Gather more computational data that will allow you to formulate hypotheses about that existence of the limits

$$\lim_{n\to\infty} D(p_n),$$

$$\limsup_{n\to\infty} D(p_n),$$

$$\liminf_{n\to\infty} D(p_n).$$

On a likely related topic, estimate the size of the initial segment of primes that is included in the closure $\overline{\{2,p\}}$.

Here are some steps taken in a MAPLE environment to work with the function $D(p_n)$. Let's plan to use **DENSITY(N)** for $D(p_n)$, since the notation **D** is reserved by MAPLE.

We may start by introducing the magma operation "$*$":

```
P:=n->max(1,max(op(factorset(n))));

f:=(x,y)->P(x+y);
```

We then initialize the original set consisting of 2 and $p_n$; let's say, for example, that $n = 27$:

```
G:={2,ithprime(27)}; N:=numelems(G); K:=N^2;
```

The reason for using $K = N^2$ here is that we plan to use the following bijection from $\{1, 2, \ldots, N^2\}$ to $\{1, 2, \ldots, N\} \times \{1, 2, \ldots, N\}$, which follows from the division algorithm:

$$x \mapsto \left(1 + \left\lfloor \frac{x-1}{N} \right\rfloor, 1 + (x-1) \bmod N\right).$$

This allows us to rewrite a nested loop of the general type

> *for I from 1 to N do*
>> *for J from 1 to N do*
>>> *[execute something with I, J]*
>> *end do*
> *end do*

as a single loop:

$$for\ k\ from\ 1\ to\ N^2\ do\ I := 1 + \left\lfloor \frac{k-1}{N} \right\rfloor; J :$$
$$= 1 + (k-1)\ \mathrm{mod}\ N; [execute\ something\ with\ I, J]\ end\ do$$

Now let's take the first step by enlarging the initial 2-element set $G$ consisting of 2 and $p_n = p_{27} = 103$ by including all possible products $x * y$ among the terms of $G$. At the end, we reset the values for $N$ and $K$ so that we can just hit enter again to move one step further in the set iteration:

```
SET:={}: for s from 1 to K do SET:=SET union {f(G[1+((s-1) mod
N)],G[1+floor((s-1)/N)])} end do: G:=G union SET;
N:=numelems(G); K:=N^2;
```

This particular line leads to updating $G$ from $G = G_0 = \{2, 103\}$ to

$$G_1 = G_0 \cup \{x * y | x, y \in G_0\} = \{2, 7, 103\},$$
$$G_2 = G_1 \cup \{x * y | x, y \in G_1\} = \{2, 3, 7, 11, 103\},$$
$$G_3 = G_2 \cup \{x * y | x, y \in G_2\} = \{2, 3, 5, 7, 11, 13, 19, 53, 103\}.$$

This is followed, using the same generation rule, by

$$G_4 = \{2, 3, 5, 7, 11, 13, 19, 29, 53, 61, 103\},$$
$$G_5 = \{2, 3, 5, 7, 11, 13, 17, 19, 29, 31, 37, 41, 53, 61, 103\},$$
$$G_6 = \{2, 3, 5, 7, 11, 13, 17, 19, 23, 29, 31, 37, 41, 43, 47, 53, 61, 67, 103\},$$
$$G_7 = \{2, 3, 5, 7, 11, 13, 17, 19, 23, 29, 31, 37, 41, 43, 47, 53, 61, 67, 73, 103\},$$
$$G_8 = \{2, 3, 5, 7, 11, 13, 17, 19, 23, 29, 31, 37, 41, 43, 47, 53, 61, 67, 73, 103\}.$$

That $G_7 = G_8$ (and consequently $G_7 = G_n$ for every $n \geq 7$) indicates a stop in the process of calculating the magma closure, which has the following explicit form:

$$\overline{G} = \bigcup_{r \geq 0} G_r = G_7.$$

Note that in this case, the size of the initial segment of primes that is included in the closure $\overline{\{2, 103\}}$ is 16.

To summarize, we repeat the process as long as the size of $G$ increases. Note that since every size increase is at least 1, it follows that to find the closure of $\{2, p_{27}\}$ it will be *sufficient* to apply the updating step $n$ (equal to 27 in this example) times. This is, however, not *necessary*: in practice, we may get to the closure $\overline{G}$ earlier. Based on this idea, the following MAPLE line iterates $G_r \mapsto G_r \cup \{x * y | x, y \in G_r\}$ and continues for as long as the number of elements in $G_r$ increases on updating.

```
LOW:=0; for r from 1 to n while numelems(G)>LOW do
N:=numelems(G); K:=N^2;  LOW:=numelems(G): for s from 1 to K do
G:=G union {f(G[1+((s-1) mod N)],G[1+floor((s-1)/N)])} end do:
end do: G; numelems(G);
```

Therefore, combining all of this into a single MAPLE procedure for the "DENSITY(n)" function yields the following code:

```
DENSITY:=proc(n::integer)
G:={2,ithprime(n)};
LOW:=0; N:=numelems(G); K:=N^2;
for v from 1 to n while numelems(G)>LOW do LOW:=numelems(G): SET:={}:
for r from 1 to K do SET:=SET union {f(G[1+((r-1) mod
N)],G[1+floor((r-1)/N)])} end do: G:=G union SET; N:=numelems(G);
K:=N^2; end do:
DENSITY:=evalf(numelems(G)/n);
end proc;
```

At this point we should be ready to go, being able to calculate and visualize directly:

```
DENSITY(27);
```

$$0.7407407407$$

```
L:=[seq(DENSITY(n),n=5..27)];
```

L := [1.200000000, .8333333333, 1.142857143, .8750000000, .6666666667, 1.100000000,
.8181818182, .5833333333, 1.076923077, .8571428571, .4000000000, 1., .9411764706,
.5555555556, .6842105263, 1.050000000, .6666666667, .7272727273, .7391304348,
.9166666667, .4800000000, .8846153846, .7407407407]

```
listplot(L);
```

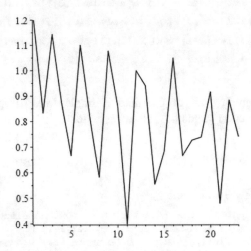

NOTE. Since getting visuals is always good and inspiring, we can use MAPLE to build 2D matrix plots for the ∗ operation on some submagmas. Of course, we have to define the operation first (with the greatest prime factor $P$ introduced previously).

```
f:=(m,n)->P(ithprime(m)+ithprime(n));
A:=matrix(31, 31, f): matrixplot(A);
```

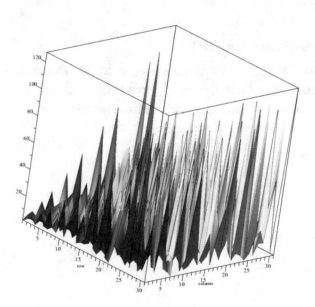

There are several plot options under MAPLE's matrixplot. Try them and enjoy the results.

## 3.5   Solvability: A Surprising Property of a Class of Infinite-Order GPF Recurrences

The experimentation with finite-order greatest prime factor sequences satisfying

$$x_n = P(c_0 + c_1 x_{n-1} + \cdots + c_d x_{n-d}) \quad \text{for} \quad n \geq d$$

led us to the plausibility of the GPF conjecture in all cases, with proofs concluded in a few special ones (including the class of second-order GPF-Fibonacci sequences).

However, we could not find a way to foresee (at least with the elementary methods that we used) the structure of the limit cycle, given $c_i$'s and the $d$ initial conditions.

Also, with the exception of first-order recurrences of the special form $x_n = P(x_{n-1} + b)$ we could not find bounds for the location of the limit cycle in terms of the recurrence parameters indicated above. That is why, at least for a while, we decided to resuscitate an age-old rule of herding: *seek greener pastures.*

So in this particular case, we decided to stop a little and take a look at what might happen with *infinite-order* greatest prime factor recurrences, in a joint work (Caragiu et al. 2012) between our group at Ohio Northern and Alexandru Zaharescu, of the University of Illinois. The one infinite-order recurrence relation that naturally came to us in this context was the following:

$$x_n = P(x_1 + x_2 + \cdots + x_{n-1}) \quad \text{for} \quad n \geq 2. \tag{3.24}$$

Here, everything proceeds from an arbitrarily chosen prime "seed" $x_1$ as initial condition.

The infinite-order recurrence (3.24) can be resolved if we transform it into a finite-order recurrence that is "mixed" in the sense that its right-hand side contains both terms that are the greatest prime factor of an expression that depends on the preceding $d$ terms, i.e., terms of the form $P(E(x_{n-1}, \ldots, x_{n-d}))$, and terms that are an expression that depends on the preceding $d$ terms without incorporating the function $P$, i.e., terms of the form $F(x_{n-1}, \ldots, x_{n-d})$. For example, a simple such mixed greatest prime factor recurrence relation is

$$x_n = P(E(x_{n-1}, \ldots, x_{n-d})) + F(x_{n-1}, \ldots, x_{n-d}).$$

To do this transformation, let us rewrite (3.24) using the cumulative sums

$$s_n := x_1 + x_2 + \cdots + x_n.$$

Since $x_n = s_n - s_{n-1}$, the recurrence (3.24) can be written as a mixed GPF recurrence as follows:

$$s_n = s_{n-1} + P(s_{n-1}) \quad \text{for} \quad n \geq 2. \tag{3.25}$$

Note that the seed $s_1$ for (3.25) is the same as the seed for (3.24), since $s_1 = x_1$. Let's get some "eyes" by experimenting with this process.

We will start by choosing $s_1 = 2$, so that we may use the following straightforward MAPLE sequence generation to get a plot of the first $N$ terms of the corresponding sequence $\{x_n\}_{n \geq 1}$.

```
s(1):=2: for n from 2 to N do s(n):=s(n-1)+P(s(n-1)) end do:
S:=[seq(s(i),i=1..N)]: X:=[seq(s(i+1)-s(i),i=1..N-1)]:
listplot(X);
```

Here is the image we get for $N = 1000$. The plot evolves, with small and seemingly random variations, in the immediate neighborhood of the line $y = x/2$.

The picture is typical for all values of $N$ tried, suggesting the asymptotic behavior

$$x_n \approx \frac{n}{2}.$$

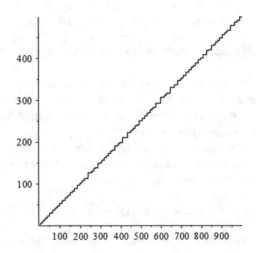

The first 100 terms (listed below) display a relatively steady and slow growth marked by groups of repeated primes:

2, 2, 2, 3, 3, 3, 5, 5, 5, 5, 7, 7, 7, 7, 7, 7, 11, 11, 11, 11, 11, 11, 13, 13, 13, 13,

13, 13, 17, 17, 17, 17, 17, 17, 19, 19, 19, 19, 19, 19, 23, 23, 23, 23, 23, 23, 23,

23, 23, 23, 29, 29, 29, 29, 29, 29, 29, 29, 31, 31, 31, 31, 31, 31, 31, 31, 37,

37, 37, 37, 37, 37, 37, 37, 37, 37, 41, 41, 41, 41, 41, 41, 43, 43, 43, 43, 43,

43, 47, 47, 47, 47, 47, 47, 47, 47, 47, 47, 53, 53

We see that each prime is present in the list, and they appear in repeated blocks, with the number of occurrences for 2, 3, 5, 7, 11, 13, ... being 3, 3, 4, 6, 6, 6, ... respectively.

Figuring out the pattern is a rewarding exercise. At least it was for us when we first started to look into this problem.

On deciphering this pattern, its simplicity led us to calling this infinite-order GPF sequence "solvable."

Going back to the picture, even if we change the initial condition $x_1$, the asymptotic behavior would be the same, even if it began with a different pattern (e.g., constant for a while). Here is what we get with the seed $x_1 = 431$:

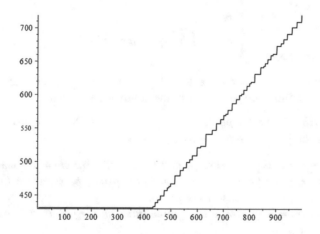

Now that we know that $x_n \approx \frac{n}{2}$, a natural step will be to see what happens with the sequence of differences $x_n - \frac{n}{2}$. Again, MAPLE will help. The following list plot displays terms of the sequence $\left\{ x_n - \frac{n}{2} \right\}_{n=2}^{1,000,000}$ for $x_1 = 2$:

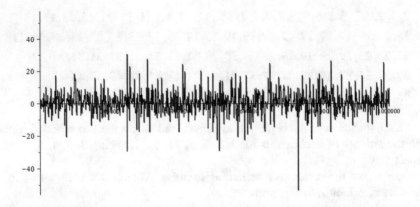

We will now proceed to show that these sequences are explicitly solvable, with their terms expressible in terms of the sequence of primes, regardless of the choice of the seed $x_1$. Let us agree to call the sequence resulting from $x_1 = 2$ the "prototype" sequence, which will be analyzed first:

$$\begin{cases} x_1 = 2, \\ x_n = P\left(\sum_{i=1}^{n-1} x_i\right) & \text{for} \quad n \geq 2. \end{cases}$$

The computer experiments conducted suggest specific regularities of the prototype sequence that are spelled out in the following theorem.

**Theorem 3.18** *The prototype sequence* $\{x_n\}_{n\geq 1}$ *and the associated sequence of partial sums* $\{s_n\}_{n\geq 1}$ *have the following properties:*

- *For every* $r \geq 1$, *the* $r$th *prime* $p_r$ *is a term of the prototype sequence.*
- *For every* $r \geq 2$, *all occurrences of* $p_r$ *in the prototype sequence form a single block of* $p_{r+1} - p_{r-1}$ *consecutive terms.*
- *The prime* $p_1 = 2$ *appears exactly three times, in the first three positions of the prototype sequence.*
- *The prototype sequence is nondecreasing.*
- *The sequence of partial sums* $\{s_n\}_{n\geq 1}$ *satisfies* $s_{p_n+p_{n+1}-2} = p_n p_{n+1}$ *for every* $n \geq 1$.
- *The sequence of partial sums consists of successive arithmetic progressions in that for every* $j$ *with* $0 \leq j \leq p_{n+2} - p_n$, *the terms* $s_{p_n+p_{n+1}-2+j}$ *can be explicitly determined as*

$$s_{p_n+p_{n+1}-2+j} = p_n p_{n+1} + j p_{n+1}, \tag{3.26}$$

*while the associated constant block of* $\{x_n\}_{n\geq 1}$ *is determined by* $x_{p_n+p_{n+1}-2+j} = p_{n+1}$ *for* $1 \leq j \leq p_{n+2} - p_n$.

*Proof of Theorem 3.18*  The key component lies in proving (3.26). Before starting the proof, just to have a more concrete idea about the arithmetic progressions mentioned, here are the first few terms of $\{s_n\}_{n\geq 1}$ (we introduced, for technical reasons, a conventional extra term $s_0 := 0$ at the beginning, standing for the empty sum):

$$0, 2, 4, 6, 9, 12, 15, 20, 25, 30, 35, 42, 49, 56, 63, 70, 77, 88, 99, 110, 121,$$
$$132, 143, 156, 169, 182, 195, 208, 221, 238, 255, 272, 289, 306, 323, 342,$$
$$361, 380, 399, 418, 437, 460, 483, 506, 529, 552, \ldots.$$

We will prove by induction on $n \geq 1$ that $s_{p_n+p_{n+1}-2+j} = p_n p_{n+1} + j p_{n+1}$ is true for $0 \leq j \leq p_{n+2} - p_n$. As the basis step, first consider $n = 1$, in which case we need to show that $s_{2+3-2+j} = p_1 p_2 + j p_2$, or $s_{j+3} = 6 + 3j$ for $0 \leq j \leq 3(=p_3 - p_1)$, which is easily verified. Let us assume, for a given $n \geq 1$, that $s_{p_n+p_{n+1}-2+j} = p_n p_{n+1} + j p_{n+1}$ for $0 \leq j \leq p_{n+2} - p_n$. We need to show that

$$s_{p_{n+1}+p_{n+2}-2+j} = p_{n+1} p_{n+2} + j p_{n+2} \quad \text{for} 0 \leq j \leq p_{n+3} - p_{n+1}. \tag{3.27}$$

Note that for $j = p_{n+2} - p_n$, (3.26) becomes

$$s_{p_{n+1}+p_{n+2}-2} = p_n p_{n+1} + p_{n+1}(p_{n+2} - p_n) = p_{n+1} p_{n+2} + 0 \cdot p_{n+2}.$$

At this moment, an increasing segment of $\{s_n\}_{n\geq 1}$ is initiated that is an arithmetic progression of difference $p_{n+2}$. To see this, let's begin by using the recurrence relation satisfied by $\{s_n\}_{n\geq 1}$ in which every term equals the previous term plus the greatest prime factor of the previous term. We get

$$s_{(p_{n+1}+p_{n+2}-2)+1} = s_{p_{n+1}+p_{n+2}-1} = s_{p_{n+1}+p_{n+2}-2} + P\big(s_{p_{n+1}+p_{n+2}-2}\big)$$
$$= p_{n+1} p_{n+2} + P(p_{n+1} p_{n+2})$$
$$= p_{n+2}(p_{n+1} + 1) = p_{n+1} p_{n+2} + p_{n+2}.$$

This will be followed by

$$s_{(p_{n+1}+p_{n+2}-2)+2} = s_{p_{n+1}+p_{n+2}-1} + P\big(s_{p_{n+1}+p_{n+2}-1}\big)$$
$$= p_{n+1} p_{n+2} + p_{n+2} + P(p_{n+1} p_{n+2} + p_{n+2})$$
$$= p_{n+1} p_{n+2} + p_{n+2} + p_{n+2} = p_{n+1} p_{n+2} + 2 p_{n+2}.$$

and then by

$$s_{(p_{n+1}+p_{n+2}-2)+3} = p_{n+1} p_{n+2} + 2 p_{n+2} + P(p_{n+1} p_{n+2} + 2 p_{n+2})$$
$$= p_{n+1} p_{n+2} + 2 p_{n+2} + p_{n+2} = p_{n+1} p_{n+2} + 3 p_{n+2}.$$

This goes on further in increments of $p_{n+2}$ (as long as the $P$ term evaluates as $p_{n+2}$). The last step for which this temporary pattern holds is at

$$s_{(p_{n+1}+p_{n+2}-2)+(p_{n+3}-p_{n+1})}$$
$$= p_{n+1}p_{n+2} + (p_{n+3} - p_{n+1} - 1)p_{n+2} + P(p_{n+1}p_{n+2} + (p_{n+3} - p_{n+1} - 1))$$
$$= p_{n+2}p_{n+3} - p_{n+2} + P(p_{n+2}p_{n+3} - p_{n+2})$$
$$= \begin{cases} p_{n+1}p_{n+2} + (p_{n+3} - p_{n+1})p_{n+2} \text{ (ending series of increments with } p_{n+2}), \\ p_{n+2}p_{n+3} + 0 \cdot p_{n+3} \text{ (inaugurating series of increments with } p_{n+3}). \end{cases}$$

Thus (3.27) is proved true for $0 \le j \le p_{n+3} - p_{n+1}$. This concludes the induction proof of (3.26). Note that the fact that the first three terms satisfy $s_1 = s_2 = s_3 = 2$ can be easily verified. Everything else in Theorem 3.18 follows from the complete description of $\{s_n\}_{n \ge 1}$ together with $x_n = s_n - s_{n-1}$. Indeed, for $0 \le j \le p_{n+2} - p_n$ we have

$$x_{p_n + p_{n+1} - 2 + j} = s_{p_n + p_{n+1} - 2 + j} - s_{p_n + p_{n+1} - 2 + j - 1}$$
$$= p_n p_{n+1} + j p_{n+1} - (p_n p_{n+1} + (j-1)p_{n+1}) = p_{n+1}.$$

The next result will discuss the asymptotic behavior of the prototype sequence $\{x_n\}_{n \ge 1}$.

**Proposition 3.19** *The prototype sequence satisfies*

$$\lim_{n \to \infty} \frac{x_n}{n} = \frac{1}{2}.$$

*Proof of Proposition 3.19* The integer sequence $\{p_k + p_{k+1} - 2\}_{k \ge 1}$ is strictly increasing. Therefore for every $n \ge 3$, there exists exactly one $k \ge 1$ with the property

$$p_k + p_{k+1} - 2 \le n < p_{k+1} + p_{k+2} - 2. \tag{3.28}$$

The above relation allows us to write $n$ in the form
$$n = p_k + p_{k+1} - 2 + j$$

for some $j$ with $0 \le j < p_{k+2} - p_k$. Theorem 3.18 combined with (3.28) leads to the following estimate for $\frac{x_n}{n} = \frac{p_{k+1}}{n}$:

$$\frac{p_{k+1}}{p_{k+1} + p_{k+2} - 2} \le \frac{x_n}{n} \le \frac{p_{k+1}}{p_k + p_{k+1} - 2}. \tag{3.29}$$

But the prime number theorem in the form $p_k \sim k \ln k$ $(k \to \infty)$ shows that

$$\lim_{k \to \infty} \frac{p_{k+1}}{p_{k+1} + p_{k+2} - 2} = \lim_{k \to \infty} \frac{p_{k+1}}{p_k + p_{k+1} - 2} = \frac{1}{2}. \tag{3.30}$$

From (3.29) and (3.30) we get the desired result $\lim_{n \to \infty} \frac{x_n}{n} = \frac{1}{2}$.

### 3.5.1 The Case of an Arbitrary Seed

From this point on we will consider sequences $\{u_n\}_{n\geq 1}$ with an arbitrary seed $u_1 = M \in \mathbb{N}$ satisfying the same recurrence relation as the one satisfied by the prototype $\{x_n\}_{n\geq 1}$, i.e.,

$$u_n = P(u_1 + u_2 + \cdots + u_{n-1}) \quad \text{for} \quad n \geq 2.$$

We will find the explicit structure of such GPF sequences in three steps. The first one establishes a powerful shift relation with the prototype, albeit conditional.

**Proposition 3.20** *With the above notation, let*

$$\sigma_n := u_1 + u_2 + \cdots + u_n.$$

*Assume that for some positive integers $L, k$ we have*

$$\sigma_L = p_k p_{k+1}. \tag{3.31}$$

*Then there exists an integer $T$ such that the following relation between $\{u_n\}_{n\geq 1}$ and the prototype $\{x_n\}_{n\geq 1}$ holds:*

$$u_n = x_{n+T}$$

*for all sufficiently large $n$.*

*Proof of Proposition 3.20* From (3.31) and Theorem 3.18 (using the notation therein) we have

$$s_{p_k + p_{k+1} - 2} = \sigma_L.$$

Since the sequences $\{u_n\}_{n\geq 1}$ and $\{x_n\}_{n\geq 1}$ satisfy the same GPF recurrence with every term equaling the greatest prime factor of the cumulative sum of the preceding terms, the above equality can be rephrased by saying that the cumulative sum of the first $p_k + p_{k+1} - 2$ terms of $\{x_n\}_{n\geq 1}$ equals the cumulative sum of the first $L$ terms of $\{u_n\}_{n\geq 1}$, so by the nature of the recurrence, it follows that

$$u_{L+r} = x_{p_k + p_{k+1} - 2 + r}$$

for all $r \geq 1$. This concludes the proof of Proposition 3.20.

The next theorem is an important result, showing that the premise of the above conditional result actually happens. Actually we enjoyed trying it by hand, and we would recommend it to everybody.

Example: let's say we start the mixed recurrence $\sigma_n = \sigma_{n-1} + P(\sigma_{n-1})$ of $\{u_n\}_{n\geq 1}$ with the seed $\sigma_1 = u_1 = 33$. The terms of $\{\sigma_n\}_{n\geq 1}$ will evolve as follows, until we get to one of them in the desired form $\sigma_{11} = 143 = 11 \cdot 13 = p_5 p_6$.

| $n$ | $\sigma_n$ | $P(\sigma_n)$ |
|-----|-----------|---------------|
| 1   | 33        | 11            |
| 2   | 33 + 11 = 44 | 11         |
| 3   | 44 + 11 = 55 | 11         |
| ... | ...       | ...           |
| 9   | 110 + 11 = 121 | 11       |
| 10  | 121 + 11 = 132 | 11       |
| 11  | 132 + 11 = 143 | 13       |
| 12  | 143 + 13 = 156 | 13       |
| ... | ...       | ...           |

**Theorem 3.21** *With the above notation, for some positive integers $L, k$ the equality $\sigma_L = p_k p_{k+1}$ is valid.*

*Proof of Theorem 3.21* If $u_1(= \sigma_1) = 1$, the verification is straightforward. The sequence of partial sums satisfying $\sigma_n = \sigma_{n-1} + P(\sigma_{n-1})$ proceeds as $1, 2, 4, 6, 9, 12, 15, 20, \ldots$, so $\sigma_L = p_k p_{k+1}$ occurs for $L = 4, 7, \ldots$.

Let us now assume $u_1(= \sigma_1) \geq 2$.

We will use the following notation convention for $n \geq 2$:

$$\sigma_{n-1} = u_n v_n, \tag{3.32}$$

where $u_n = P(\sigma_{n-1})$ and $v_n = \sigma_{n-1}/P(\sigma_{n-1}) = \sigma_{n-1}/u_n$. In terms of $u_n$ (the "prime part" of $\sigma_{n-1}$ in the above decomposition) and $v_n$, the recurrence satisfied by $\{\sigma_n\}$ can be rewritten in the following form, which holds for $n \geq 2$:

$$u_{n+1} v_{n+1} = \sigma_n = \sigma_{n-1} + P(\sigma_{n-1}) = u_n v_n + P(u_n v_n) = u_n v_n + u_n = u_n(v_n + 1). \tag{3.33}$$

Let's try to understand the type of changes undergone by $u_n$ and $v_n$ at each step. From (3.33) it follows that the $u$ and $v$ components of the products (3.32), having the essential property that $u_n$ is prime and $P(v_n) \leq u_n$, will change under the following transition rules when moving from $u_n v_n$ to $u_{n+1} v_{n+1}: P(v_n + 1)$:

**(T1)** If $P(v_n + 1) \leq u_n$ then $u_{n+1} = u_n$ and $v_{n+1} = v_n + 1$.
**(T2)** If $P(v_n + 1) > u_n$ then $u_{n+1} = P(v_n + 1)$ and $v_{n+1} = u_n(v_n + 1)/P(v_n + 1)$.

This makes the sequence $\{u_n\}$ nondecreasing, with an infinite limit.

We will now prove an auxiliary result to the effect that if $v_n \leq u_n$ for some $n \geq 2$ then the conclusion of Theorem 4 holds.

Indeed, let's say that

$$v_n \leq u_n = p_k.$$

If $p_k = 2$ then either $\sigma_{n-1} = u_n v_n = 2$ (in which case $\sigma_{n+1} = 2 \cdot 3$), or $\sigma_{n-1} = u_n v_n = 4$ (in which case $\sigma_n = 2 \cdot 3$).

Thus we may assume $p_k \geq 3$, with $v_n \leq u_n = p_k$. Then (T1) repeatedly applies, with the $v$ component increasing in steps of 1 until it reaches the value $p_{k+1} - 1$. That's exactly the moment when (T2) applies, making the $u$ component equal to $p_{k+1}$, while the $v$ component equals $p_k$. In other words, for some $m > n$ we have $u_m = p_{k+1}, v_m = p_k$ and consequently, with $L := m+1$ we have $\sigma_L = p_k p_{k+1}$. This proves the auxiliary result.

Getting back to the proof of Theorem 3.21, the above auxiliary result gives us the road to the proof's conclusion. The only thing we need to do now is to rule out the possibility of the series of inequalities

$$u_n < v_n \quad \text{for all} \quad n \geq 2. \tag{3.34}$$

In this context, let's take a closer look at the behavior of $u_n, v_n$ under iteration:

- If the transition rule (T1) applies, the $v$ component *increases by* 1 (in particular, preserving the validity of (3.34), since the $u$ component does not change).
- If the transition rule (T2) applies (which means that $P(v_n + 1) > u_n$), the $v$ component does not increase; indeed, $v_{n+1} = u_n(v_n + 1)/P(v_n + 1) \leq v_n$.

Therefore, at each step, if the $v$ component increases, it can increase only in steps of 1. On the other hand, (3.34) together with the fact that $\{u_n\}$ has an infinite limit implies that the sequence $\{v_n\}$ is not bounded from above. Thus for some $n$, the component $v_n$ will be a prime, say $v_n = p_k$ and $u_n = p_l$ ($u_n$ is always prime) and $u_n = p_l < p_k = v_n$ by (3.34). On the other hand,

$$p_1 = u_n = P(\sigma_{n-1}) = P(p_k p_l) = p_k,$$

in contradiction to $p_l < p_k$. Thus the proof of Theorem 3.21 is complete.

The following theorem is a direct consequence of Proposition 3.20 and Theorem 3.21, to the effect that the arbitrary seed case of the recurrence relation discussed here can be eventually be written as a shift of the prototype sequence $\{x_n\}_{n \geq 1}$ with seed equaling 2.

**Theorem 3.22** *Let $\{u_n\}_{n \geq 1}$ be a recurrent sequence satisfying*

$$\begin{cases} u_1 = M \in \mathbb{N} \\ u_n = P(u_1 + u_2 + \cdots + u_{n-1}) & \text{for} \quad n \geq 2 \end{cases} \tag{3.35}$$

Then

$$\lim_{n \to \infty} \frac{u_n}{n} = \frac{1}{2}.$$

Using a result on primes in small intervals, we can provide a sharper asymptotic form for these GPF sequences.

**Theorem 3.23** *Every GPF sequence defined by a recurrence as in* (3.35) *can be asymptotically estimated as*

$$u_n = \frac{n}{2} + O\left(n^{0.525}\right). \tag{3.36}$$

*Proof of Theorem 3.23* Since a sequence $\{u_n\}_{n \geq 1}$ satisfying (3.35) is related to the prototype $\{x_n\}_{n \geq 1}$ (i.e., (3.35) with $M = 2$) by a relation of the form $u_n = x_{n+T}$ for all large enough $n$, it will be enough to prove (3.36) for the prototype sequence.

To do this, we will use the following heavy result from analytic number theory, which provides a bound on gaps between consecutive primes.

**Lemma 3.24** Baker et al. (2001) *For all sufficiently large $x$ there exists a prime in the interval $\left[x, x + x^{0.525}\right]$.*

From Lemma 3.24, it follows that

$$p_k = p_{k+1} + O\left(p_{k+1}^{0.525}\right),$$

and

$$p_{k+2} = p_{k+1} + O\left(p_{k+1}^{0.525}\right).$$

Using the above two estimates in (3.29) with the notation therein, it follows that

$$\frac{x_n}{n} = \frac{p_{k+1}}{n} = \frac{p_{k+1}}{2p_{k+1} + O\left(p_{k+1}^{0.525}\right)} = \frac{1}{2}\left(1 + O\left(\frac{1}{p_{k+1}^{0.475}}\right)\right).$$

On the other hand, substituting $p_{k+1} = O(n)$ in the above relation, we get

$$\frac{x_n}{n} = \frac{1}{2}\left(1 + O\left(\frac{1}{n^{0.475}}\right)\right).$$

This completes the proof of Theorem 3.23.

We will conclude here the analysis, following (Caragiu et al. 2012), of the sequences (3.35). To summarize, we derived a simple explicit form in terms of primes of the $M = 2$ case (the "prototype sequence"), we proved that every other seed eventually leads to a shift of the prototype, and we derived a sharp asymptotic estimate that holds for all sequences in this class.

So what now? One alternative is to experiment some more with similar infinite-order GPF recurrences. Here is a potentially interesting exploration project that I would like to share.

**Computational Exploration Project CEP9** Investigate infinite-order greatest prime factor sequences $\{u_n\}_{n \geq 1}$ satisfying a recurrence of the following form

$$\begin{cases} u_1 = M \in \mathbb{N}, \\ u_n = P(u_{n-1} + 2u_{n-2} + 2u_{n-3} + \cdots + 2u_1) \quad \text{for} \quad n \geq 2. \end{cases} \quad (3.37)$$

More generally, choose various sequences $c_1, c_2, c_3, \ldots$ of positive integers and explore the GPF recurrences of the form $u_n = P(c_1 u_{n-1} + c_2 u_{n-2} + c_3 u_{n-3} + \cdots + c_{n-1} u_1)$.

Let us quickly discuss how to get started on (3.37) with MAPLE. Imitating the model of (3.25), we can rewrite (3.37) as a "mixed" GPF recurrence in terms of partial sums $s_n = u_1 + u_2 + \cdots + u_n$. Note that $u_{n-1} + 2u_{n-2} + 2u_{n-3} + \cdots + 2u_1 = s_{n-1} + 2s_{n-2}$, so that (3.37) becomes

$$s_n = s_{n-1} + P(s_{n-1} + s_{n-2}). \quad (3.38)$$

This is a second order recurrence with initial conditions $s_1 = u_1$ and $s_2 = u_1 + u_2 = u_1 + P(u_1)$, since $u_2 = P(u_1)$. Conversely, the sequence $\{u_n\}_{n \geq 1}$ can be recovered by $u_n = s_n - s_{n-1}$ for $n \geq 2$ (here we can extend it to $n \geq 1$ by adopting the standard empty-sum convention $s_0 := 0$).

Let us address the case $u_1 = 2$ (although this time, we will see no strong reasons to call this a "prototype" sequence). MAPLE can be used to find a number of initial terms:

```
N:=50: s(1):=2: s(2):=4: for k from 3 to N do s(k):=s(k-
1)+P(s(k-1)+s(k-2)) end do: S:=[seq(s(k),k=1..N)]: X:=[seq(s(k)-
s(k-1),k=2..N)];
```

2, 2, 3, 11, 5, 41, 29, 157, 7, 13, 31, 571, 23, 31, 607, 2459, 17, 127, 181, 107, 8741, 41, 26371,

52783, 199, 26417, 42307, 40037, 757, 17539, 223, 29297, 37, 389, 11083, 8951, 176747, 769,

178691, 3637, 114113, 2339, 297889, 7681, 7151, 7673, 757, 193939, 251, 7529

There is no clearly identifiable pattern, at least not with so few terms. But the problem seems quite interesting, especially when we use MAPLE as before to visualize the logarithmic plot of the first 2000 terms of $\{u_n\}_{n \geq 1}$.

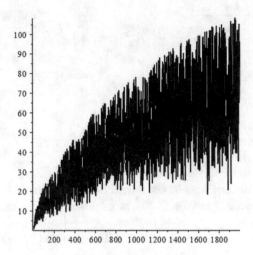

Do we detect any overall pattern now? Any conjectures? Let us hope that continuing with CEP9 will lead to pleasant surprises.

## 3.6  GPF Ducci Games: A Combinatorial Unleashing of 2, 3, 5, 7

The classical Ducci games (see Section 2.6) are based on the fact that the Ducci iteration (operating on $n$-tuples of nonnegative integers) defined by

$$D(x_1, x_2, x_3, \ldots, x_{n-1}, x_n) = (|x_1 - x_2|, |x_2 - x_3|, |x_3 - x_4|, \ldots, |x_{n-1} - x_n|, |x_n - x_1|)$$

ultimately leads to a limit cycle, which is of length 1 if $n$ is a power of 2. Moreover, the components of the vectors in the limit cycle are all in a set $\{0, m\}$ (which makes the cycle iteration essentially binary, since $|x - y| \equiv x + y \pmod{2}$ for integers $x, y$.

We will construct an alternative of the Ducci iteration using the greatest prime factor function. Of course, a viable alternative should preserve the following two essential features:

- a result on ultimate periodicity;
- a result restricting the components of the vectors in the cycle.

Our GPF Ducci games were introduced in Caragiu et al. (2014). *The d-number GPF Ducci games are played on the set $\Pi^d$ of d-tuples of primes* and are obtained by replacing the "local interaction" function $(x, y) \mapsto |x - y|$ with the GPF analogue $(x, y) \mapsto P(x + y)$, where $P$ is the greatest prime factor function. Thus the GPF

Ducci iteration $G : \Pi^d \to \Pi^d$ is defined as follows:

$$G(x_1, x_2, x_3, \ldots, x_{d-1}, x_d) = (P(x_1 + x_2), P(x_2 + x_3),$$
$$P(x_3 + x_4), \ldots, P(x_{d-1} + x_d), P(x_d + x_1)). \quad (3.39)$$

As a numerical example, the table below illustrates the GPF Ducci iteration starting with the initial vector seed $(11, 17, 2, 13, 7, 19)$. In that case, the iteration enters a cycle of length 12.

| 11 | 17 | 2 | 13 | 7 | 19 |
|---|---|---|---|---|---|
| 7 | 19 | 5 | 5 | 13 | 5 |
| 13 | 3 | 5 | 3 | 3 | 3 |
| 2 | 2 | 2 | 3 | 3 | 2 |
| 2 | 2 | 5 | 3 | 5 | 2 |
| 2 | 7 | 2 | 2 | 7 | 2 |
| 3 | 3 | 2 | 3 | 3 | 2 |
| 2 | 5 | 5 | 3 | 5 | 5 |
| 7 | 5 | 2 | 2 | 5 | 7 |
| 3 | 7 | 2 | 7 | 3 | 7 |
| 5 | 3 | 3 | 5 | 5 | 5 |
| 2 | 3 | 2 | 5 | 5 | 5 |
| 5 | 5 | 7 | 5 | 5 | 7 |
| 5 | 3 | 3 | 5 | 3 | 3 |
| 2 | 3 | 2 | 2 | 3 | 2 |
| 5 | 5 | 2 | 5 | 5 | 2 |
| 5 | 7 | 7 | 5 | 7 | 7 |
| 3 | 7 | 3 | 3 | 7 | 3 |
| 5 | 5 | 3 | 5 | 5 | 3 |
| 5 | 2 | 2 | 5 | 2 | 2 |
| 7 | 2 | 7 | 7 | 2 | 7 |
| 3 | 3 | 7 | 3 | 3 | 7 |
| 3 | 5 | 5 | 3 | 5 | 5 |
| 2 | 5 | 2 | 2 | 5 | 2 |
| 7 | 7 | 2 | 7 | 7 | 2 |
| 7 | 3 | 3 | 7 | 3 | 3 |
| 5 | 3 | 5 | 5 | 3 | 5 |
| 2 | 2 | 5 | 2 | 2 | 5 |
| 2 | 7 | 7 | 2 | 7 | 7 |
| 3 | 7 | 3 | 3 | 7 | 3 |
| 5 | 5 | 3 | 5 | 5 | 3 |
| ... | ... | ... | ... | ... | ... |

To get a quick feeling for how we can proceed with a computational analysis, it's very easy to use MATLAB. Recall that the greatest prime factor function can be obtained as

```
function gpf = gpf(n);
gpf=max(factor(n));
end
```

To perform *n* GPF Ducci iterations starting from an initial vector seed *x*, we can use the following MATLAB function:

```
function gpfducci = gpfducci(x,n);
k=size(x,2);
for m=1:k;
a(1,m)=x(m);
end;
for I=2:n;
for J=1:k-1;
a(I,J)=gpf(a(I-1,J)+a(I-1,J+1));
end;
a(I,k)=gpf(a(I-1,k)+a(I-1,1));
end;
gpfducci=a;
```

We have now to choose a seed before we can begin playing the game. For example, let's say that we want to start with a random initial prime vector. We can get such a vector fairly quickly using a simple MATLAB code **rprimes(d)** to generate a vector of length *d* whose components are small (less than 100) primes:

```
function [rprimes]=rprimes(n)
for I=1:n;
x(I)=ithprime(floor(100*rand()));
end;
rprimes=x;
```

Now let's use these to play for a while, say a 15-number GPF Ducci game, and visualize its first 10 steps.

```
≫ x = rprimes(15)
x =
  349    5  103    7   23  421  347  127  499    5  191  163  383  401   61
≫ gpfducci(x,10)
ans =
349     5  103    7   23  421  347  127  499    5  191  163  383  401   61
 59     3   11    5   37    3   79  313    7    7   59   13    7   11   41
 31     7    2    7    5   41    7    5    7   11    3    5    3   13    5
 19     3    3    3   23    3    3    3    3    7    2    2    2    3    3
 11     3    3   13   13    3    3    3    5    3    2    2    5    3   11
  7     3    2   13    2    3    3    2    2    5    2    7    2    7   11
```

| 5 | 5 | 5 | 5 | 5 | 3 | 5 | 2 | 7 | 7 | 3 | 3 | 3 | 3 | 3 |
| 5 | 5 | 5 | 5 | 2 | 2 | 7 | 3 | 7 | 5 | 3 | 3 | 3 | 3 | 2 |
| 5 | 5 | 5 | 7 | 2 | 3 | 5 | 5 | 3 | 2 | 3 | 3 | 3 | 5 | 7 |
| 5 | 5 | 3 | 3 | 5 | 2 | 5 | 2 | 5 | 5 | 3 | 3 | 2 | 3 | 3 |

After the game has been played repeatedly, both desired features appear to be true: we notice that most of the time, the iterative flow (3.39) leads to a vector in which all components form an old acquaintance: the four-element magma $\{2, 3, 5, 7\}$ (see Section 3.4).

Since $\{2, 3, 5, 7\}$ is invariant under the local GPF Ducci interaction term $(x, y) \mapsto P(x+y)$, once we get to a vector with all components in $\{2, 3, 5, 7\}$, all subsequent vectors will have components in $\{2, 3, 5, 7\}$, and the ultimate periodicity follows, since from that point on, the vectors in the GPF Ducci iteration will be confined to a finite set.

We considered this to be an excellent candidate for an analogue of the classical Ducci games that is played with primes and uses the greatest prime factor function.

The main result proved in Caragiu et al. (2014) is the following:

- The iteration of $G : \Pi^d \to \Pi^d$ defined by (3.39) always leads into a limit cycle $C$.
- If $C$ is nontrivial (has length greater than 1), then every vector in $C$ has components in $\{2, 3, 5, 7\}$

Conversely, if $C$ is nontrivial, then every element of $\{2, 3, 5, 7\}$ appears as a component of some vector in the limit cycle $C$.

The actual proof of ultimate periodicity for the iteration (3.39) on $\Pi^d$ is actually very simple, and it makes a lot of sense after Section 3.4, which deals with GPF magmas.

**Theorem 3.25** *Every iteration on $\Pi^d$ defined by the recurrence (3.39) is ultimately periodic.*

*Proof of Theorem 3.25* Assume that the seed $x \in \Pi^d$ is $x = (x_1, x_2, \ldots, x_d)$ with $x_i \in \Pi$ for $i = 1, 2, \ldots, d$. Let us choose a prime $p \equiv 1 \pmod 3$ such that $p \geq x_i$ for every $i = 1, 2, \ldots, d$. Since every component of the seed $x = (x_1, x_2, \ldots, x_d)$ is an element in the submagma (closed under the binary operation $*$, "greatest prime factor of the sum")

$$K_p := \{x | x \text{ prime}, x \leq p\},$$

it follows that every component of every subsequent vector in the iteration, e.g., $G(x)$, $G(G(x))$, ..., is in $K_p$. Since all the vectors in the iteration have components in the finite set $K_p$, the ultimate periodicity of this first-order vector iteration follows.

Of course, there are also trivial cycles consisting of a single prime vector, a $d$-tuple whose components are necessarily equal and do not need to be among the first four primes. There are nonconstant $d$-tuples that are mapped into a constant d-tuple via the $G$-iteration; for example, if $d$ is even, this is the case of a $d$-tuple of the form $(p, q, p, q, p, q, \ldots, p, q)$, although more complex situations may occur, such as $(5, 17, 71, 61) \longrightarrow [G](11, 11, 11, 11)$. If we restrict, for example, to a seed with components in $K_7 = \{2, 3, 5, 7\}$, the probabilities $pr_d$ of getting into a trivial cycle on following the $d$-number GPF Ducci game with $d = 3, 4, 5, 6, 7, 8$ are given in the following table:

| $d$ | $pr_d$ |
|---|---|
| 3 | 0.062500 |
| 4 | 0.906250 |
| 5 | 0.003906 |
| 6 | 0.293945 |
| 7 | 0.000244 |
| 8 | 0.297973 |

In any case, the odds appear markedly different in the even and odd cases, quickly approaching zero if the dimensionality of the game is odd.

We already know from Theorem 3.25 that every such sequence of GPF Ducci iterates is ultimately periodic. Therefore, for every seed $x \in \Pi^d$ there exist $L = L_x$ and $n_0 = n_x$ such that for all $n \geq n_0$, one has

$$G^{n+L}(x) = G^n(x).$$

The following fact is remarkable: for the classical Ducci game, the iteration is essentially binary once we get in the cycle, while for the GPF Ducci analogues, the corresponding calculations in the iteration are done with "GPF addition" in the four-element nonassociative magma $K_7$. In this sense, the novelty of our construction is also marked by a transition from the binary realm to a nonassociative quaternary one.

Here is the main result of ultimate periodicity (Caragiu et al. 2014).

**Theorem 3.26** *Let $x \in \Pi^d$ with $L_x > 1$. Then for all $n \geq n_x$, the nth iterate $G^n(x)$ has all components in $K_7 = \{2, 3, 5, 7\}$. Moreover, each element of $K_7$ appears as component in one of the vectors in the limit cycle, that is, it appears as a component of one of the prime vectors $G^n(x)$ for $n \geq n_x$.*

Once the above theorem is proved, we will provide computational data pertaining to dimensions $d = 3, 4, 5, 6, 7, 8$, showing counts of all possible limit cycles and their numbers of occurrences if one starts from a seed $x \in K_7^d$.

The proof is a case-by-case analysis of elementary number-theoretic arguments, with a flavor of combinatorial number theory. In fact, I believe that the proper place of the GPF Ducci games in the mathematical classification is largely under "combinatorial number theory."

For every $k$ with $1 \leq k \leq L = L_x$, let us define

$$G^{n_0 + k - 1}(x) = (p_{k,1}, p_{k,2}, \ldots, p_{k,d}).$$

Once we get in the limit cycle, the game will be perpetually confined to an $L \times d$ matrix with prime numbers as entries. Let's call it $A$. This is a very important matrix, since the limit cycle $C$ of the GPF Ducci iteration consists of the rows of $A$.

The $k$th row of $A$ will be $G^{n_0 + k - 1}(x)$. Since it was assumed that the cycle is nontrivial, it follows that $L > 1$ and the vector $(p_{1,1}, p_{1,2}, \ldots, p_{1,d})$ is nonconstant, i.e., not a constant multiple of $(1, 1, 1, \ldots, 1)$.

*Proof of Theorem 3.26* The proof will be carried out in four steps in the form of lemmas.

**Lemma 3.27** Let $q$ be the largest entry in the matrix $A$. Then $q$ is odd, and if the primes $a$ and $b$ are consecutive entries in a row of $A$ that generate $q = P(a + b)$ in the next row, then one of the following must hold:

- $a = b = q$.
- The element $q - 2$ is prime, and either $(a, b) = (2, q - 2)$ or $(a, b) = (q - 2, 2)$.

*Proof of Lemma 3.27* We begin with visual aid for what will follow:

| ... | $a$ | $b$ | ... |
|-----|-----|-----|-----|
| ... | $q$ | ... | ... |

Indeed, $q$ must be odd, for otherwise, the largest element of the cycle matrix $A$ is 2, which would mean $L_x = 1$, i.e., that the limit cycle is trivial.

Assume that it is not the case that $a = b = q$. This means that either $a \neq q$ or $b \neq q$.

We can quickly rule out the cases in which one of $a, b$ equals $q$ while the other does not. For example, if $a \neq q$ and $b = q$, then $P(a + b) = P(a + q)$ cannot possibly be equal to $q$, since otherwise we would have $q | a$, implying $q = a$, in contradiction to the hypothesis $a \neq q$. The same type of argument shows that the case $a = q$ and $b \neq q$ cannot occur.

Next we will rule out the case of odd $a, b$ with $a \neq q$ and $b \neq q$. Since $q$ is maximal in $A$, the previous two inequalities actually read $a, b < q$ with $a, b$ odd. But then $q = P(a + b) \leq (a + b)/2 < q$, a contradiction.

Therefore, either $a = 2$ or $b = 2$. Recalling that $P(a + b) = q$, let us assume, without loss of generality that the case $q = 2$ holds (and of course $b$ is a prime less than $q$, from the previous reasoning). Then the element $2 + b$ must be prime. Otherwise, we would have $q = P(a + b) = P(2 + b) < 2 + b \leq q$, a contradiction. Thus $2 + b$ is prime, and this prime necessarily equals $q$, so $a + 2$ and $b = q - 2$. A similar argument can be made in the case $b = 2$, in which case we can repeat the previous derivation to show that $a = q - 2$. This concludes the proof of Lemma 3.27.

**Lemma 3.28** Let $p := q - 2$. Then $p$ appears in the cycle matrix $A$.

*Proof of Lemma 3.28* Assume that $p$ does not appear as an entry of $A$. Then the maximum entry $q$ (which does appear in $A$) appears precisely in places where there are consecutive occurrences of $q$ in the previous row:

| ... | $q$ | $q$ | ... |
|-----|-----|-----|-----|
| ... | $q$ | ... | ... |

Together with the periodic boundary conditions imposed on $A$ by cyclicity, this implies that $q$ appears in every row of $A$, and an application of the iteration rule $G$ produces a strictly smaller number of $q$ entries in the next row, which is impossible (the function *number of the occurrences of q in the nth row* would then be periodic and decreasing). Thus $p$ must appear in the cycle matrix $A$.

**Lemma 3.29** The largest entry $q$ of the cycle matrix $A$ is at most 7.

*Proof of Lemma 3.29* We proceed by contradiction. Let us assume that $q \geq 11$. Then $p$ (which appears in $A$) is odd, and $p - 2 = q - 4$ is composite. Let the primes $a$ and $b$ be two neighboring entries in a row of $A$ such that in the next row they will generate $P(a+b) = p$. We will show that both $a$ and $b$ must be equal to $p$ by showing that every possible option will lead to the relation $P(a+b) \neq p$, which contradicts the hypothesis made above.

(i) If $a \neq p$ and $b = p$, since $p$ does not divide the prime $a$, it follows that $P(a+b) \neq p$. The case $a = p$ and $b \neq p$ is treated similarly.

(ii) Let $a,b$ be both odd and $p \notin \{a, b\}$. If $a, b$ are both smaller than $p$, then due to $a+b$ being even, we have $P(a+b) \leq (a+b)/2 < p$. If $a,b$ are both greater than $p$, then due to the fact that $p + 2 = q$ with $q$ maximal, it follows that $a = b = q$, in which case $P(a+b) \neq p$. If $a < p$ and $b > p$, then $b = q$ and $a < p - 2$ (since $p - 2$ is composite), in which case $P(a+b) \leq (a+b)/2 < (p - 2 + q)/2 = p$. The same reasoning works for $b < p$ and $a > p$.

(iii) If $a = b = 2$, then $P(a+b) = 2 \neq p$ ($p$ is odd).

(iv) Let $a = 2$ and let $b$ be an odd prime. Since $p - 2$ is composite, $b \neq p - 2$, so $p \neq 2 + b = a + b$. Here we distinguish two possibilities. If $2 + b$ is prime, then $P(a+b) = P(2+b) = 2 + b \neq p$. On the other hand, if $2 + b$ is composite, then being composite and odd implies

$$P(a+b) = P(2+b) < \frac{2+b}{3} \leq \frac{2+q}{3} = \frac{2+b+2}{3} = \frac{4+b}{3} < p,$$

so $P(a+b) \neq p$.

This proves that if $P(a+b) = p$, then $a = b = p$. Thus every occurrence of $p$ in the cycle matrix corresponds to two consecutive occurrences of $p$ in the previous row (in the "above" and "above-right" positions). Together with the fact that $p$ does appear in the cycle matrix (Lemma 3.28) and the cyclic structure of $A$, $p$ will appear in every row of $A$. Also, since $p - 2$ is composite, it cannot appear in $A$, so that as in the proof of Lemma 3.28, the function *number of occurrences of p in the nth row* is strictly decreasing, which is inconsistent with the cyclic character of $A$. Therefore, the maximal element of $A$ satisfies $q \leq 7$. This concludes the proof of Lemma 3.29.

Note an important consequence: we got to a point where we know that every entry of the cycle matrix $A$ is in $K_7 = \{2, 3, 5, 7\}$ or in terms of the iteration mapping, $G^n(x) \in K_7^d$ for $n \geq n_x$.

**Lemma 3.30** *Each of the primes 2, 3, 5, and 7 is an entry of the matrix A.*

*Proof of Lemma 3.30* The proof will proceed in a number of steps, each of them offering a clearer view of the cycle matrix $A$.

*First step: if 3 appears in some row of A, then 5 must appear in A.*

Indeed, if 3 appears in a row of $A$, then that row has entries other than 3 (otherwise, we would have a trivial cycle). If there is a 5 in the row, then we are done. Otherwise, some 3 in that row must appear next to either a 7 or a 2, in which case 5 will appear in the next row.

*Second step: if 7 appears in some row of A, then 3 must appear in A.*

Indeed, if 7 appears in a row of $A$, then that row has entries other than 7. If there is a 3 in this row, then there is nothing to prove. Otherwise, if 3 is missing, then somewhere in that row we will see a 7 next to a 5, or a 7 next to a 2. In either case, 3 will appear in the next row.

*Third step: if 7 appears in some row of A, then 5 and 3 must appear in A.*

This comes naturally as a consequence of the first two steps.

*Fourth step: if 7, 5, and 3 appear in A, then 2 must appear in A.*

To accomplish this, we will show that 3 and 5 appear as nearest neighbors in some row of $A$. Consider a row of $A$ containing a 3, and assume that 2 is not in this row (otherwise there is nothing to prove). If 5 does not appear next to that 3, then somewhere in that row we must see one of the strings 337, 733, 737. From the first two strings we get either 35 or 53 in the next row. From the segment 737 we get 55 in the next row. Thus we either get 557 or 755 if there is no 3 next to 5 in $A$. But then in the subsequent row we do get 53 or 35. Therefore, if 7 appears in some row of $A$, then all three numbers 5, 3, and 2 must appear in $A$. The proof of Lemma 4 (and hence of Theorem 3.1) will be completed if we prove that 7 appears in $A$.

*Fifth step:* 7 *appears in A.*

Suppose the element 7 is missing from $A$. Then 2 and 5 can't appear next to each other in a row, since 7 will be immediately generated in the next row. If 2 and 5 appear in a row separated by 3's, then it is easy to see that a repeated application of $G$ to a particular segment of the form 233 …. 35 or 533 … 32 ultimately produces a 7. Therefore *we can't have both* 2 *and* 5 *in the same row.*

If a (necessarily nonconstant) row contains both 3's and 5's (therefore no 2's and no 7's), then 3 and 5 must appear next to each other somewhere in that row. Then, since 7 is not an entry of $A$, a segment of the form 35 or 53 may only be produced, via $G$, from a segment of the form 332 or 233 in the previous row, which in turn may only be a subsegment of one of 3322, 3323, 2233, 3233 in the same row. In the first and third cases we get 2 and 5 after applying $G$ once. In the second and fourth cases we get 2 and 5 after applying $G$ twice. Hence, 3 and 5 can't appear in the same row.

Lastly, the possibility of a nonconstant row consisting of 2 and 3 only may be ruled out, since that would imply that the previous row contains both 3 and 5 (only the segments 3 5 and 5 3 may generate a 2 via $G$).

To summarize, if 7 does not appear in $A$ (i.e., each entry of $A$ is 2, 3, or 5) and if $L_x > 1$, we have proved that

- 2 and 5 cannot appear in the same row.
- 3 and 5 cannot appear in the same row.
- There is no row consisting of 2's and 3's only.

This contradiction shows that if the limit cycle is nontrivial, then 7 must appear in $A$. This concludes the proof of Lemma 3.30, and hence, together with the previous three lemmas, the proof of the important Theorem 3.26.

### Computational Data for $d$-Number GPF Ducci Games, $3 \leq d \leq 8$

The tables below (Caragiu et al. 2014) summarize the distribution of the limit cycle lengths for the GPF Ducci games of dimensions from 3 to 7, with all possible cycle lengths and examples of seeds leading to a nontrivial cycle of a given length $L$. As an interesting fact, the distribution of cycle lengths in even dimensions appears to favor smaller cycle lengths, while for odd dimensions, it is tilted toward the larger cycle lengths.

- 3-number GPF Ducci games

Distribution of GPF Ducci cycle lengths for seeds $x \in K_7^3$

| $L$ | # Seeds $x \in K_7^3$ with limit period $L$ | Seed leading to a nontrivial cycle of length $L$ |
|-----|----------------------------------------------|--------------------------------------------------|
| 1   | 4                                            |                                                  |
| 12  | 60                                           | (2,3,5)                                          |

- 4-number GPF Ducci games

Distribution of GPF Ducci cycle lengths for seeds $x \in K_7^4$

| $L$ | # Seeds $x \in K_7^4$ with limit period $L$ | Seed leading to a nontrivial cycle of length $L$ |
|---|---|---|
| 1 | 232 | |
| 2 | 4 | (2,7,3,5) |
| 4 | 4 | (5,3,7,2) |
| 8 | 16 | (7,2,3,5) |

- 5-number GPF Ducci games

Distribution of GPF Ducci cycle lengths for seeds $x \in K_7^5$

| $L$ | # Seeds $x \in K_7^5$ with limit period $L$ | Seed leading to a nontrivial cycle of length $L$ |
|---|---|---|
| 1 | 4 | |
| 20 | 70 | (5,3,3,7,7) |
| 30 | 120 | (5,7,2,5,5) |
| 40 | 830 | (7,2,5,7,7) |

- 6-number GPF Ducci games

Distribution of GPF Ducci cycle lengths for seeds $x \in K_7^6$

| $L$ | # Seeds $x \in K_7^6$ with limit period $L$ | Seed leading to a nontrivial cycle of length $L$ |
|---|---|---|
| 1 | 1204 | |
| 6 | 1428 | (2,3,3,5,3,3) |
| 12 | 1116 | (5,2,3,5,2,2) |
| 27 | 174 | (5,7,3,2,7,3) |
| 54 | 174 | (3,3,7,2,5,2) |

- 7-number GPF Ducci games

Distribution of GPF Ducci cycle lengths for seeds $x \in K_7^7$

| $L$ | # Seeds $x \in K_7^7$ with limit period $L$ | Seed leading to a nontrivial cycle of length $L$ |
|---|---|---|
| 1 | 4 | |
| 4 | 196 | (5,7,3,3,3,7,7) |
| 28 | 196 | (7,7,7,5,7,5,5) |
| 126 | 3528 | (3,5,7,7,7,5,7) |
| 168 | 12,460 | (5,3,2,2,7,3,2) |

- 8-number GPF Ducci games

Distribution of GPF Ducci cycle lengths for seeds $x \in K_7^8$

| $L$ | # Seeds $x \in K_7^8$ with limit period $L$ | Seed leading to a nontrivial cycle of length $L$ |
|---|---|---|
| 1 | 19,528 | |
| 2 | 4 | (2,7,3,5,2,7,3,5) |
| 4 | 4 | (2,5,3,7,2,5,3,7) |
| 8 | 12,496 | (3,5,2,7,7,3,7,5) |
| 16 | 14,808 | (7,5,5,5,7,3,5,3) |
| 32 | 14,816 | (5,7,7,7,3,7,2,3) |
| 48 | 240 | (3,7,3,7,7,3,2,5) |
| 80 | 3640 | (7,5,5,7,5,2,3,3) |

Among the MATLAB codes that we designed to deal with GPF Ducci periods, besides the function **gpfducci(x)** presented previously, we present some of the codes that were used in our semiempirical computational work on the subject. The first few programs illustrate some elementary steps we took to work with vectors with components in $K_7$. To generate a random element in $K_7^n$:

```
function randomvf = randomvf(n);
for I=1:n;
x(I)=ithprime(1+mod(floor(100*rand(1)),4));
end;
randomvf=x;
```

To generate all vectors in $K_7^n$:

```
function gen=gen(n);
p=4^n;
q=4^(n+1)-1;
for I=p:q;
a=digits(I,4);
for k=1:n;
x(I,k)=ithprime(a(k)+1);
end;
end;
for I=1:p-1;
x(1,:)=[];
end;
gen=unique(x,'rows');
```

Here **digits(I,4)** produces a vector with the digits of $I$ to base 4. The code used was as follows:

```
function digits= digits(n,b);
q=n;
I=1;
x(1)=mod(q,b);
while q~=0;
    q=(q-mod(q,b))/b;
    I=I+1;
    x(I)=mod(q,b);
end;
k=size(x,2);
for J=1:k-1;
    y(J)=x(J);
end;
digits=y;
```

For **ithprime(k)**, or the $k$th prime. (In this very brute force setup, $k$ ranges up to 78,498. However, if necessary we can decrease the range of the **primes** function inside):

```
function ithprime = ithprime(n)
x=primes(1000000);
ithprime=x(n);
```

To find the GPF Ducci period, if we start with a seed vector $x \in \Pi^n$:

```
function gpfducciperiod - gpfducciperiod(x,r)
y=gpfducci(x,r);
for I=1:floor(r/2);
y(1,:)=[];
end;
u=seqperiod(y);
gpfducciperiod=u(1);
```

This was a little bit on the empirical side in a joint enterprise between MATLAB and human gut feeling, but it was monitored extensively in dimensions up to 8, so that all possible periods are covered and their frequencies accounted for. Starting with $x \in \Pi^n$ (in the work done, we considered only $n \leq 8$) we generate the first $r$ rows in the GPF Ducci game (we used $r = 3000$, clearly overkill, but it did the job), remove the first half of its rows so that the remaining part fell into a "pure periodicity" regime, and then applied to it the known MATLAB function **seqperiod(x)**.

For the analysis of the Ducci periods of games played in $K_7^n$:

```
function analysis = analysis(n)
a=gen(n);
for I=1:4^n;
x(I)=period(gpfducci(a(I,:),3000));
end;
histc(x,unique(x))
analysis=distinct(x);
```

This is expected to produce a description of the GPF Ducci cycle lengths with the number of times a cycle of any produced length is reached in terms of the chosen seed. For example, the input $n = 3$ produces

```
≫ analysis(3)
ans =
    4    60
ans =
    1    12
```

That means out of $4^3 = 64$ possible seeds in $K_7^3$, four of them lead to a cycle of length 1 (clearly these are the four constant vectors), while 60 of them enter a cycle of length 12. Of course, the running time of this function increased considerably for the next five values of the argument. At $n = 8$, I had to leave the office, ensure that the computer would not go to sleep, and return the next day to see what was up.

### 3.6.1 GPF Ducci Period Fishing with a Monte Carlo Rod

For a Monte Carlo–type randomized exploration or the lengths of the limit cycles:

```
function search = search(searchspace,vectorsize);
for I=1:searchspace;
a=randomvf(vectorsize);
for J=1:vectorsize;
m(I,J)=a(J);
end;
x(I)=gpfducciperiod(a,1000);
end;
[frequencies,periodlength]=hist(x,unique(x))
```

Obviously, the number of trials should be adjusted to a reasonably large value in order to "see" more periods. For example, it is possible that 200 random searches would produce all periods of the 4-number GPF Ducci game:

```
≫ search(200,4)
frequencies =
   182   1   3   14
periodlength =
    1    2   4    8
```

However, it happened that going with only 100 random searches left one of the periods out.

I am pretty sure that through more sophisticated programming this could be improved, and it would have made the process less time-consuming, but for the moment, I figured that the tradeoff for the excitement of number-theoretic data made this worthwhile, mathematical ideas being paramount.

Note that the GPF Ducci iteration can be placed in the context of multidimensional GPF sequences (MGPF, Section 3.3). If we represent the vectors in the $d$-number GPF Ducci games as column vectors $Q_n \in K_7^d$, then the iteration can be represented in vector form as $Q_n = P(\Lambda Q_{n-1})$ for $n \geq 1$, where $Q$ is the $d \times d$ circulant matrix with first row $(1100\ldots0)$.

### 3.6.2 An Infinite-Dimensional Analogue

What about GPF recurrences that iterate not vectors of a fixed length $d$, but infinite sequences with prime components? The GPF Ducci iteration was designed by replacing the "local interaction" $(x, y) \mapsto P(x + y)$ used in the classical $d$-number Ducci game with $(x, y) \mapsto P(x + y)$ at the cost of playing the game with $d$-tuples with prime components. If we think about infinite sequences, the following iteration was the one used in formulating the Proth–Gilbreath conjecture (see Section 2.6):

$$(x_1, x_2, \ldots, x_{n-1}, x_n, \ldots) \mapsto (|x_1 - x_2|, |x_2 - x_3|, \ldots, |x_{n-1} - x|, |x_n - x_{n+1}|, \ldots).$$

The Proth–Gilbreath conjecture began with the sequence of primes and asserted that repeatedly applying the above iteration would produce 1 as the first entry of each of the subsequent rows. But since, as one may argue, the GPF Ducci iteration format seems a natural choice as a transition rule between sequences of primes, why shouldn't we try to see what happens if we consider a greatest prime factor analogue of the Proth–Gilbreath conjecture? This means that starting from a "seed" represented by the sequence of primes, we repeatedly apply the iteration where the local interaction is $(x, y) \mapsto P(x + y)$:

$$(x_1, x_2, \ldots, x_{n-1}, x_n, \ldots) \mapsto (P(x_1 + x_2), P(x_2 + x_3), \ldots, P(x_{n-1} + x_n), (x_n + x_{n+1}), \ldots).$$

$$(3.40)$$

To our satisfaction, this was done in Caragiu et al. (2013), where we managed to prove the following GPF analogue of the Proth–Gilbreath conjecture.

**Theorem 3.31** *Caragiu et al. (2013) Assume that the iteration (3.40) is repeatedly applied starting from the "seed" represented by the sequence of primes* $(p_1, p_2, p_3, \ldots, p_{n-1}, p_n, \ldots)$. *Let the k-th vector in the iteration be denoted by* $(p_{k,1}, p_{k,2}, p_{k,3}, \ldots, p_{k,n-1}, p_{k,n}, \ldots)$. *Then*

$$p_{k,1} \in \{2, 3, 5, 7\}$$

*for every* $k \geq 1$.

Illustrating this result is the table below, representing the northwestern corner of the infinite 2D array that has $\{p_{k,n}\}_{n \geq 1}$ as the $k$th row.

| 2 | 3 | 5 | 7 | 11 | 13 | 17 | 19 | 23 | 29 | 31 | ... |
|---|---|---|---|----|----|----|----|----|----|----|-----|
| 5 | 2 | 3 | 3 | 3 | 5 | 3 | 7 | 13 | 5 | 17 | ... |
| 7 | 5 | 3 | 3 | 2 | 2 | 5 | 5 | 3 | 11 | 5 | ... |
| 3 | 2 | 3 | 5 | 2 | 7 | 5 | 2 | 7 | 2 | 5 | ... |
| 5 | 5 | 2 | 7 | 3 | 3 | 7 | 3 | 3 | 7 | 7 | ... |
| 5 | 7 | 3 | 5 | 3 | 5 | 5 | 3 | 5 | 7 | 3 | ... |
| 3 | 5 | 2 | 2 | 2 | 5 | 2 | 2 | 3 | 5 | 5 | ... |
| 2 | 7 | 2 | 2 | 7 | 7 | 2 | 5 | 2 | 5 | 2 | ... |
| 3 | 3 | 2 | 3 | 7 | 3 | 7 | 7 | 7 | 7 | 7 | ... |
| 3 | 5 | 5 | 5 | 5 | 5 | 7 | 7 | 7 | 7 | 3 | ... |
| 2 | 5 | 5 | 5 | 5 | 3 | 7 | 7 | 7 | 5 | 2 | ... |
| ... | ... | ... | ... | ... | ... | ... | ... | ... | ... | ... | ... |

The proof is similar in style to that of the ultimate periodicity of the GPF Ducci sequences. The essential structure introduced in the proof was that of "admissible path" starting from an element in the first column. In the above grid, let us call $p_{k-1,n}$ and $p_{k-1,n+1}$ the "left parent" and "right parent," respectively, of the $k$th row element $p_{k,n}$. By an "admissible path" of a prime number $q$ in the $n$th row we understand a sequence $q = q_1, q_2, \ldots, q_n$ such that for each $j \geq 2$ the element $q_j$ is the larger of the two parents of $q_{j-1}$, unless $q_{j-1}$ has equal parents, in which case $q_j$ is the left parent of $q_{j-1}$. The proof takes a close look at the admissible paths originating with an element $q = q_1$ of the first column, and proceeds through a series of key steps that are instrumental in the derivation. We outline them below, and refer the reader to Caragiu et al. (2013) for details.

- *Lemma on turns in admissible paths*: if $1 \leq j \leq n - 1$ then $q_{j+1} \geq q_j$ unless $q_j$ and $q_{j+1} = q_j - 2$ form a twin prime pair.
- *Lemma on the structure of admissible paths*: if $j \leq n - 2$ satisfies $q_j \geq 11$ then in the admissible path through $q_j$ either $q_{j+1} \geq q_j$ or there exists $k > j$ such that $q_l = q_j - 2$ for $j < l < k$ and at the same time $q_l > q_j$ for all $l \geq k$. Moreover, if $q_{j+1} > q_j$ then $q_{j+1} > q_j + 2$. Consequently, any increase in such an admissible path must be by an amount $d > 2$ and therefore cannot be overturned by a subsequent decrease.
- *Closing in*: Assuming that an admissible path starts from $q_1 \geq 11$, then all its subsequent terms are at least 11.
- *Final step*: assume for the sake of a contradiction that the originating first-column element in an admissible path satisfies $q_1 \geq 11 = p_5$. Let $q_n = p_k$ (which means that the admissible path originating at $q_1$ has $k - 1$ right turns). First, note that $q_j \geq 11$ for all $j$ with $1 \leq j \leq n$. Walking along the admissible path, at each right turn from $q_j$ we either move to a bigger prime (which subsequently leads to a terminal segment consisting of entries greater than $q_j$) or decreases by 2 (moving on to $q_{j+1} = q_j - 2$), an amount strictly less than the size of any increase that may previously have occurred. Therefore, each of the $k - 1$ right turns generates a new prime, and therefore, the admissible path through $q_1 \geq p_5$ covers at least $k$ primes along its way. The last element, $q_n$, is either the largest of all the $q_j$'s or 2 less than the largest of all the $q_j$'s. Therefore, $p_k = q_n \geq p_{5+(k-1)-1} = p_{k+3}$, a contradiction, showing that the first column origin of the admissible path satisfies $q_1 < 11$, thus proving this analogue of the Proth–Gilbreath conjecture.

## A Comparative Analysis

We now proceed with a computational comparative analysis of the convergence rates in classical versus GPF Proth–Gilbreath iterates. This analysis is interesting, since it suggests a conjectural generalization of the theorem we just proved.

Let us assume that we allow both types of iterates (classical Proth–Gilbreath $\sigma_{PG}$ and its GPF analogue $\sigma_{GPF}$) to be applied starting from a sufficiently long vector representing an initial segment of consecutive primes

$$\Pi_M := (p_1, p_2, \ldots, p_M). \tag{3.41}$$

After $k$ iterates with $k < M$, (3.41) becomes $\sigma_{PG}^k(\Pi_M)$ in the classical version and $\sigma_{GPF}^k(\Pi_M)$.

Clearly every iteration (in both versions) leads to a decrease (in this finite setup) by 1 of the number of components, so that for each $k < M$ the vectors $\sigma_{PG}^k(\Pi_M)$ and $\sigma_{GPF}^k(\Pi_M)$ have $M - k$ components each. Let $X_k = \max(\sigma_{PG}^k(\Pi_M))$ and $Y_k = \max(\sigma_{GPF}^k(\Pi_M))$ respectively. We noticed that both types of iterates eventually enter a "stable regime" characterized by

- $X_k \leq 2$ in the classical Proth–Gilbreath iteration, the "PG-stable" regime,
- $Y_k \in \{2, 3, 5, 7\}$ for the GPF-analogue, the "GPF-stable" regime.

Our main "experimental" finding was the following: in the case of the GPF-analogue, the iterates enter much faster the stable regime than in the case of classical Proth–Gilbreath iterates. In our analysis (Caragiu et al. 2013) we used $M = 200,000$. Thus initially, we have

$$X_0 = Y_0 = p_{200,000} = 2,750,159$$

The following are the first few terms of the sequences of maximum values $\{X_n\}_{n \geq 0}$ and $\{Y_n\}_{n \geq 0}$, respectively.

- For the classical Proth–Gilbreath iteration, $\{X_n\}_{n \geq 0}$ starts as 2750159, 148, 142, 142, 142, 130, 130, 126, 124, 118, 112, 108, 108, 104, 104, 104, 104, 104, 102, ..., slowly moving toward the PG-stable zone, where it arrives after 113 applications of $\sigma_{PG}$: $X_{113} = 2$.
- For the GPF analogue, $\{X_n\}_{n \geq 0}$ starts as 2750159, 1374617, 1127051, 530197, 212579, 34673, 10853, 4969, 757, 211, 193, 61, 43, 23, 13, 13, 7, thus entering the "stable regime" after 16 applications of $\sigma_{GPF}$. The vector $\sigma_{GPF}^{16}(\Pi_{200,000})$ has 199,984 components, all of them in the mysterious set $K_7 = \{2, 3, 5, 7\}$.

The above computation provides supporting evidence for a generalization of the result proved in Theorem 3.31:

**PROTH–GILBREATH–GPF CONJECTURE:** For every $k \geq 1$ the GPF analogue of the Proth–Gilbreath iteration brings the components of the first $k$ columns into the set $\{2, 3, 5, 7\}$ after a sufficient number of iterations.

Indeed, the progression toward $K_7$ appears to be *very fast*.

**Computational Exploration Project CEP10** Use more computational power and/or special programming to explore in greater depth the speed of the simultaneous convergence to $K_7$ in the first $k$ columns of the GPF-analogue Proth–Gilbreath iteration. Do they all enter the set $K_7$ in $O(\ln \ln k)$ iterative steps?

## 3.7    All Primes in Terms of a Single Prime and Related Puzzles

In Section 3.4, we discussed the nonassociative magma $(\Pi, *)$, where $\Pi$ is the set of primes and $x * y = P(x+y)$. This "greatest prime factor of the sum" is of potential interest for further exploration due to its submagma structure, with $K_7 = \{2, 3, 5, 7\}$ an outstanding submagma since it is the minimal element in the ordered set of nontrivial submagmas of $(\Pi, *)$ ordered by inclusion. Also, for $p \equiv 1 \pmod 3$, the initial segment $K_p = \{x | x \text{ prime}, x \le p\}$ is a submagma, a fact that can be used to provide a quick proof that $(\Pi, *)$ is not finitely generated. This finite generation —or rather lack of it—offers the challenge of finding magma structures similar to $(\Pi, *)$ that perhaps *are* finitely generated, or even stronger, perhaps are generated by a single element! In the same section we explored the magmas $(\Pi, f_{a,b})$, which in some sense evoke Robert Musil's "man without qualities," since there is apparently nothing there for us: $(\Pi, f_{a,b})$ are nonassociative, in general noncommutative, and have no identity element. But just as the lack of qualities of Musil's hero Ulrich turned out to be itself an interesting quality, there may be more to it.

### 3.7.1    Prologue: An Exercise with Commuting Pairs

To begin, to put the structures $(\Pi, f_{a,b})$ to good use in an undergraduate classroom context, we may use them to design exercises in which students are asked to find commuting pairs.

For example, let us take $a = 2, b = 1$, and look for all nontrivial commuting pairs $(x, y)$ in the magma $(\Pi, \circ)$, where $x \circ y = P(2x+y)$, where nontrivial means $x \ne y$. That is, we have to look for all pairs of distinct primes $(x, y)$ such that

$$P(2x+y) = P(x+2y). \tag{3.42}$$

Without loss of generality, let us assume $x < y$. If $x = 2$ with $y$ odd, then (3.42) becomes

$$P(4+y) = P(2+2y) = P(y+1).$$

This necessarily leads to $P(4+y) = P(y+1) = 3$. Then $y$ must be a prime of the form $y = 3^k - 4$ for some $k \ge 2$, in which case $y+1 = 3^k - 3$ must be of the form $3 \cdot 2^T$ for some $T \ge 1$.

Therefore, we have

$$3^{k-1} - 2^T = 1. \tag{3.43}$$

That is, the difference between a power of 3 and a power of 2 is 1. It can be proved by elementary means that this can happen only for $3 - 2 = 1$ and $9 - 8 = 1$, so that all integer solutions of (3.43) are $k = 2, T = 1$ and $k = 3, T = 3$. Thus $y = 5$ or $y = 23$ with the corresponding commuting pairs $(2, 5)$ and $(2, 23)$.

It remains to consider the possibility of the case of odd primes $x, y$ satisfying (3.42). Let

$$r := P(2x + y) = P(x + 2y). \tag{3.44}$$

From (3.44) we find that in particular, $r|3(x + y)$ and $r|(x - y)$, and thus $r|6x$. But $r$ must be odd, so $r = 3$ or $r = x$. It is easy to see that the case $r = x$ implies $x = y$, so it can be ruled out. Let $P(2x + y) = P(x + 2y) = 3$. Since $x, y$ are odd, we find that for some positive integers $U, V$,

$$\begin{cases} 2x + y = 3^U, \\ 2y + x = 3^V \end{cases} \tag{3.45}$$

Since $x, y \geq 3$, we have $U, V \geq 2$. Solving the system (3.45) for $x, y$ will lead to $x = 2.3^{U-1} - 3^{V-1}$ and $y = 2.3^{V-1} - 3^{U-1}$, which means that $x, y$ are divisible by 3, and therefore $x = y = 3$, in contradiction to the nontriviality assumption. Thus, the only nontrivial commuting pairs in the magma $(\Pi, \circ)$ are $(2,5)$, $(5,2)$, $(2,23)$, $(23,2)$.

> **Computational Exploration Project CEP11** Choose a magma $(\Pi, f_{a,b})$ and use MATLAB to search for pairs of commuting elements $(x, y) \in \Pi \times \Pi$, that is, those satisfying $P(ax + by) = P(ay + bx)$. If possible, try to set up an elementary number-theoretic argument to formally derive the set of commuting pairs.

## 3.7.2   A Cyclicity Conjecture

In reference to the same structure $(\Pi, f_{2,1})$, we will now explore the possibility of the set of primes $\Pi$ being generated by a single element through recursively applying $f_{2,1}$ to pairs of elements previously obtained, following the example of the structure $(\mathbb{N}, +)$ being generated from the element 1, presented at the beginning of Section 2.5.

For $(\mathbb{N}, +)$ we advanced from the seed $A_0 = \{1\}$ by always enlarging the previous $A_i$ with the values of the operation (addition in that case) applied to all possible pairs of elements in $A_i$. Thus,

$$A_1 = \{1\} \cup \{1+1\} = \{1,2\},$$
$$A_2 = \{1,2\} \cup \{1+1,1+2,2+1,2+2\} = \{1,2,3,4\}.$$

Continuing in this manner with $A_n = \{1,2,3,\ldots,2^n\}$, we get $\mathbb{N} = \bigcup_{n=1}^{\infty} A_n$, and we say, in a fancy algebraic manner, that $\mathbb{N} = \langle 1 \rangle$, indicating a cyclic structure.

The generation growth is described by the clear-cut exponential function $|A_n| = 2^n$.

We will attempt to apply the same idea to the structure $(\Pi, \circ)$, where for notational simplicity, we will have "$\circ$" stand for $f_{2,1}$. Is there any prime that could play the role of 1 in the previous additive example? In other words, can we generate all primes in terms of a single, special one, using "$\circ$"?

A simple but important observation is that no odd prime will be able to play the role of a potential generator of $(\Pi, \circ)$. Indeed, if $x \in \Pi$ is odd, then when it comes to generating additional elements, $x$ is "dead on arrival," since

$$x \circ x = P(2x+x) = P(3x) = x.$$

On the other hand,

$$2 \circ 2 = P(2 \cdot 2 + 2) = P(6) = 3. \tag{3.46}$$

Thus our only sporting chance is with the (potential) generator $x = 2$, in which case we will be looking at the ascending sequence of sets of primes $\{G_n\}_{n \geq 0}$ with $G_0 = \{2\}$ and, for $n \geq 1$,

$$G_n = G_{n=1} \cup \{x \circ y | x, y \in G_{n=1}\}. \tag{3.47}$$

Let's try our hand first with a few iterations of (3.47). As we have seen from (3.46), we have $G_1 = \{2,3\}$. This will be followed by

$$G_2 = \{2,3\} \in \{2 \circ 3, 3 \circ 2.3 \circ 3\} = \{2,3,7\},$$
$$G_3 = \{2,3,7\} \cup \{2 \circ 7, 3 \circ 7, 7 \circ 2, 7 \circ 3\} = \{2,3,7,11,13,17\}.$$

To eliminate redundancies, consider two consecutive iterative steps

$$G_{n-2} \subset G_{n=1} \subset G_n.$$

Because the products $x \circ y$ with $x, y \in G_{n-2}$ would have been already incorporated in $G_{n-1}$, we can apply (3.47) in the *slightly* less time-consuming but equivalent form

$$G_n = G_{n-1} \cup \{x \circ y | x \in G_{n-2}, y \in G_{n-1} - G_{n-2}\} \cup \{x \circ y | x \in G_{n-1} - G_{n-2}, y \in G_{n-2}\}$$
$$\cup \{x \circ y | x, y \in G_{n-1} - G_{n-2}\}$$

The next term in the sequence $\{G_n\}_{n\geq0}$ contains every prime up to 47. Note that 5 shows up here for the first time:

$$G_4 = \{2,3,5,7,11,13,17,19,23,29,31,37,41,43,47\} \ .$$

The role of these preliminary calculations was to give an okay to initiating a process that would gather additional computational evidence in favor of the following conjecture.

**CYCLICITY CONJECTURE FOR** $(\Pi, \circ)$: The magma $(\Pi, \circ)$ is cyclic with

$$\Pi = \langle 2 \rangle = \bigcup_{n=0}^{\infty} G_n. \tag{3.48}$$

Exploring this, together with the potential cyclicity for other magmas of the form $(\Pi, f_{a,b})$ was the objective of a joint paper (Caragiu and Vicol 2016) with Paul A. Vicol, then a graduate student in computer science at Simon Fraser University.

Let us introduce some additional definitions, notation, and observations in relation to the particularities of the ascending sequence $\{G_n\}_{n\geq0}$.

- The rank $r = r(x)$ of a prime $x \in \Pi$ is the minimum nonnegative integer $r$ with $x \in G_r$ if such an integer exists; otherwise, the rank is infinite. For example, the rank of 2 is 0, the rank of 5 is 4, and the cyclicity conjecture can be rephrased as *every prime has a finite rank*. Every prime up to 137 has rank 5 or less (try to verify this by hand).
- For $n \geq 0$, denote by $R_n$ the set of primes of rank $n$. Note that the rank sets $\{R_n\}_{n\geq0}$ are disjoint. The first couple of them are $R_0 = \{2\}$, $R_1 = \{3\}$, $R_2 = \{7\}$, $R_3 = \{11,13,17\}$, and $R_4 = \{5,19,23,29,31,37,41,43,47\}$.
- Note that the cyclicity conjecture (3.48) is equivalent to $\Pi = \coprod_{n=0}^{\infty} R_n$.
- Denote by $g_n$ the number of elements in $G_n$.
- Denote by $\gamma_n$ the maximum cardinality of an initial segment of consecutive primes that is included in $G_n$. For example, from the above special cases, $\gamma_0 = 1$, $\gamma_1 = \gamma_2 = \gamma_3 = 2$, and $\gamma_4 = 15 = g_4$. Clearly, the inequality $\gamma_n \leq g_n$ holds for every $n \geq 0$.

The following table presents computational evidence for the cyclicity of $(\Pi, \circ)$ (Caragiu and Vicol 2016), displaying the constants $g_n$ and $\gamma_n$ for $0 \leq n \leq 15$ and the consecutive quotients $g_n/g_{n-1}$.

| $n$ | $g_n$ | $\gamma_n$ | $g_n/g_{n-1}$ |
|---|---|---|---|
| 0 | 1 | 1 | – |
| 1 | 2 | 2 | 2.000000 |
| 2 | 3 | 2 | 1.500000 |
| 3 | 6 | 2 | 2.000000 |
| 4 | 15 | 15 | 2.500000 |

(continued)

(continued)

| $n$ | $g_n$ | $\gamma_n$ | $g_n/g_{n-1}$ |
| --- | --- | --- | --- |
| 5 | 33 | 33 | 2.200000 |
| 6 | 76 | 73 | 2.303030 |
| 7 | 187 | 174 | 2.460526 |
| 8 | 468 | 447 | 2.502673 |
| 9 | 1223 | 1207 | 2.613247 |
| 10 | 3230 | 3178 | 2.641046 |
| 11 | 8669 | 8654 | 2.683900 |
| 12 | 23,532 | 23,450 | 2.714499 |
| 13 | 64,429 | 64,256 | 2.737931 |
| 14 | 177,643 | 177,566 | 2.757190 |
| 15 | 493,154 | 493,060 | 2.719803 |

For example, the last line communicates that all primes up to 7,259,167 (the 493,060th prime) are in the set $G_{15}$ with 493,154 elements.

It appears that the sets $G_n$ are very close to being initial segments of consecutive primes themselves.

The growth of $\{g_n\}_{n \geq 0}$ is interesting enough to warrant the following computational project.

**Computational Exploration Project CEP12** The rate of growth of $\{g_n\}_{n \geq 0}$ can be assessed in the rates of consecutive terms $g_n/g_{n-1}$, displayed in the last column. The growth appears to be exponential with a variable rate. Gather additional computational data until you are able to formulate a data-supported conjecture about the long-range behavior of the sequence $\{g_{n+1}/g_n\}_{n \geq 0}$.

The above cyclicity conjecture is equivalent to the following fact (of a "syntactic" nature):

Every prime can be written as a nonassociative product involving 2 and the parentheses ) and (.

Some examples:

$$3 = 2 \circ 2$$
$$5 = 2 \circ (2 \circ (2 \circ (2 \circ 2))),$$
$$7 = (2 \circ (2 \circ 2)),$$
$$11 = 2 \circ (2 \circ (2 \circ 2)),$$
$$13 = (2\circ)(2 \circ 2 \circ (2 \circ 2)),$$
$$17 = (2 \circ (2 \circ 2)) \circ (2 \circ 2).$$

Asking students to find such representations, even for primes that are not much larger, makes for exciting exercises.

### 3.7.3 Exploring the Possible Cyclicity of a General Magma $(\Pi, f_{a,b})$: Necessary Conditions and Computational Evidence

Let us look into the possible cyclicity of magmas $(\Pi, f_{a,b})$ by starting with the simplest case $a = b$, in which the answer is definitely negative. If $a = b = 1$, then the noncyclic nature of $(\Pi, f_{1,1})$ or $(\Pi, *)$ was settled in Section 3.4. The fact that for $a > 1$, the magma $(\Pi, f_{a,a})$ is not finitely generated and therefore cannot be cyclic will be derived in a similar way.

For the proof, first note that the binary operation $f_{a,a}$ can be expressed as

$$(x, y) \mapsto P(ax + ay) = \max(P(a), x * y).$$

Assuming that $X$ is a finite set of generators for $(\Pi, f_{a,a})$, let us choose a prime $p \equiv 1 \pmod{3}$ such that $p > \max(P(a), \max(X))$. Then the set $K_p$ of primes that are less than or equal to $p$ is closed under $f_{a,a}$. Indeed, if $x, y \leq p$, then

$$f_{a,a}(x, y) = \max(P(a), P(x + y)) = \max(P(a), x * y).$$

But $x * y \in K_p$, since $K_p$ is a submagma of $(\Pi, *)$ from the discussion in Section 3.4, and $P(a) \in K_p$ from the way we defined $p$. Hence we find that $f_{a,a}(x, y) \in K_p$. Thus we have constructed a finite submagma of $(\Pi, f_{a,b})$ that contains $X$ and therefore $\langle X \rangle \subseteq K_p \neq \Pi$. This means that $X$ cannot generate $(\Pi, f_{a,a})$, concluding the proof.

When it comes to exploring the cyclicity of $(\Pi, f_{a,b})$ in general, we decided to pursue two directions of investigation, because they were the only ones in which we believed that we could have some realistic hopes:

- Finding necessary conditions of cyclicity (possibly after going through a series of reduction steps involving the parameters $a, b$).
- For selected magmas $(\Pi, f_{a,b})$ that satisfy the necessary conditions, acquire computational evidence for cyclicity by calculating a reasonably long chain of recursively generated sets in the corresponding sequence $\{G_n\}_{n \geq 0}$.

The following simplifying assumptions (Caragiu and Vicol 2016) are warranted.

[S1]. Without loss of generality, we may assume that $a, b$ are not both even and not both divisible by the square of the same odd prime.

Indeed, if $2^k$ is the maximal power of 2 that divides both $a$ and $b$, with $a = 2^k a'$ and $b = 2^k b'$, then the magmas $(\Pi, f_{a,b})$ and $(\Pi, f_{a',b'})$ are identical, while $a', b'$ are not both even.

Indeed, this trivially results from continuing the simple step

$$f_{a,b}(x,y) = P(ax+by)P\left(2\left[\frac{a}{2}x+\frac{b}{2}y\right]\right) = \left(\frac{a}{2}x+\frac{b}{2}y\right) = f_{a/2,b/2}(x,y).$$

As an example, $(\Pi,\circ) = (\Pi,f_{2,1}) = (\Pi,f_{4,2}) = (\Pi,f_{8,4}) = \cdots$.

For the second reduction, it will be enough to show that if $r$ is an odd prime, then the magmas $(\Pi,f_{ar^2,br^2})$ and $(\Pi,f_{ar,br})$ are identical. Indeed, we have

$$f_{ar^2,br^2}(x,y) = P\left(ar^2x+br^2y\right) = P(r(arx+bry)) = \max(r,P(arx+bry)).$$

Since $P(arx+bry) \geq r$, the above relation becomes

$$f_{ar^2,br^2}(x,y) = P(arx+bry) = f_{ar,br+br}(x,y).$$

This completes the proof of the reductive step [S1].

[S2]. We may assume $a \geq b$ without loss of generality.
Indeed, if $a > b$, then $f_{a,b}(x,y) = f_{b,a}(y,x)$.
Let us agree to call $(\Pi,f_{a,b})$ *reduced* if it satisfies [S1] and [S2].

The first step in the process of finding necessary conditions for cyclicity of $(\Pi,f_{a,b})$ is finding a reasonably small "search space" where we may look for a potential generator $g$.

**Lemma 3.32** If $g$ is a generator of $(\Pi,f_{a,b})$, then $g < P(a+b)$.

*Proof of Lemma 3.32* Assume $g \geq P(a+b)$. Then

$$f_{a,b}(g,g) = P(ag+bg) = P(g(a+b)) = \max(g,P(a+b)) = g.$$

This means that starting from $g$, we cannot generate anything else, i.e., $\langle g \rangle = \{g\}$.

**Lemma 3.33** If $(\Pi,f_{a,b})$ is cyclic and $d := \gcd(a,b)$, then $P(d) \leq 3$. If, in addition, $(\Pi,f_{a,b})$ is reduced, then $P(d) \in \{1,3\}$.

*Proof of Lemma 3.33* Let $a = da_1$, $b = db_1$, and let us proceed by contradiction and assume $P(d) \geq 5$. If $x,y \in \Pi$, we have

$$f_{a,b}(x,y) = P(ax+by) = P(d(a_1x+b_1y)) = \max(P(d),P(a_1x+b_1y)) \geq P(d) \geq 5.$$
$$(3.49)$$

Let $\{G_n\}_{n \geq 0}$ be the recurrent sequence of generating sets, starting from $G_0 = \{g\}$, corresponding to the operation

$$\circ = f_{a,b}.$$

From (3.49) it follows that neither 2 nor 3 is in $G_n - G_0$ for $n \geq 1$, so that regardless of the value of $g$, either 2 or 3 will not be in $\langle g \rangle = \bigcup_{n=0}^{\infty} G_n$, and hence $g$ cannot be a generator. This proves that $P(d) \leq 3$. If $(\Pi, f_{a,b})$ is reduced, then [S1] rules out the possibility of $P(d) = 2$, and thus $P(d) \in \{1,3\}$, where $P(d) = 1$ stands for $d = 1$.

Note that Lemma 3.32 rules out the cyclicity of magmas $(\Pi, f_{a,b})$ for which $a + b$ is a power of 2. The following result (Caragiu and Vicol 2016) is an important step in the direction of establishing further restrictions on cyclicity.

For an integer $a \geq 1$, let us denote by $D_a$ the set of all prime divisors of $a$, where we agree that $D_1 = \emptyset$.

**Proposition 3.34** *If $(\Pi, f_{a,b})$ is cyclic with $\langle g \rangle = \Pi$ and $r$ is a prime divisor of $a$ that does not divide $b$, then $g = r$. In consequence, if the symmetric difference $D_a \Delta D_b$ has two or more elements, then the magma $(\Pi, f_{a,b})$ is not cyclic.*

*Proof of Proposition 3.34* Let $\langle g \rangle = \Pi$ and let $r$ be a prime with $r|a$ and $r|b$. Proceeding by contradiction, let us assume that $g \neq r$.

Let $\{G_n\}_{n \geq 0}$ be the standard recursively defined generating set pertaining to $(\Pi, f_{a,b})$ and the generator $g$, that is, $G_0 = \{g\}$ and $G_n = G_{n-1} \cup \{f_{a,b}(x,y) | x, y \in G_{n-1}\}$ for $n \geq 1$. We will prove that this leads to the following contradictory statement:

$$r \notin \bigcup_{n=0}^{\infty} G_n = \langle g \rangle = \Pi. \tag{3.50}$$

To this end, we show by induction that $r \notin G_n$ for every $n \geq 0$. Since $g \neq r$, we have $r \notin G_0 = \{g\}$. For $n \geq 1$, assume that $r \notin G_{n-1}$.

If $r \in G_n$, the form of the set recursion satisfied by $\{G_n\}_{n \geq 0}$ mandates that $r = P(ax + by)$ for some $x, y \in G_{n-1}$. Here we use the assumption $r|a$ and $r|b$ to conclude that $r|y$ and thus $r = y$, since both $r$ and $y$ are primes. But then $r \in G_{n-1}$, a contradiction, thus proving the contradictory statement (3.50). This means the hypothesis $g \neq r$ is false. Therefore, we have proved the first part of Proposition 3.34 to the effect that if $r$ is a prime divisor of $a$ that does not divide $b$, then $g = r$.

For the second part, let us assume that $D_a \Delta D_b = (D_a - D_b) \cup (D_b - D_a)$ has at least two elements. Let us now proceed by cases.

A.  If there are at least two elements in $D_a - D_b$, let $r_1, r_2 \in D_a - D_b$ with $r_1 \neq r_2$. From the now proved first part of the proposition, if the structure $(\Pi, f_{a,b})$ were cyclic with generator $g$, then $g = r_1$ and $g = r_2$ would both have to be true, a contradiction. Thus $(\Pi, f_{a,b})$ is not cyclic.

B. Similarly, if there are at least two elements in $D_a - D_b$, then proceeding as before, we again conclude that $(\Pi, f_{a,b})$ is not cyclic.

C. Finally, if $r_1 \in D_b - D_a$, $r_2 \in D_a - D_b$, and $(\Pi, f_{a,b})$ is cyclic with generator $g$, then $r_1 \in D_a - D_b$ implies $g = r_1$ and $r_2 \in D_b - D_a$ implies $g = r_2$, which is again a contradiction.

Thus, the assumption $|D_a \Delta D_b| \geq 2$ rules out the cyclicity of the structure $(\Pi, f_{a,b})$. This completes the proof of Proposition 3.34.

The following theorem is the main result obtained in Caragiu and Vicol (2016). It lists a series of fairly restrictive necessary conditions for the cyclicity of $(\Pi, f_{a,b})$. In our plan, this was the last step before proceeding to the gathering of computational evidence for selected magmas that do satisfy the necessary conditions thus derived.

**Theorem 3.35** *A magma $(\Pi, f_{a,b})$ that is cyclic and reduced, with generator $g$, necessarily falls into one of the following four categories*

- $a = 3^k, b = 3, g = 2$, with $k \geq 2$;
- $a = p^k, b = 1, g = p$, with $k \geq 1, p$ prime, $P(p^k + 1) > p$;
- $a = 3^k 2^l, b = 3, g = 2$, with $k, l, \geq 1$;
- $a = \max(3^k, 3 \cdot 2^l), b = \min(3^k, 3 \cdot 2^l), g = 2$, with $k, l, \geq 1$.

*Proof of Theorem 3.35* From Proposition 3.34, the set of primes $D_a \Delta D_b$ has either zero elements (i.e., $D_a \Delta D_b = \varnothing$) or one element. Let's discuss both of these cases.

A. Let $D_a \Delta D_b = \varnothing$. The case $D_a = D_b = \varnothing$ is easily ruled out. Then $a = b - 1$, in which case $(\Pi, f_{1,1})$ is known to be noncyclic. Otherwise, $D_a = D_b$ must be two equal sets of odd primes, since $(\Pi, f_{a,b})$ is reduced. If we consider Lemma 3.33, the only possible common odd prime divisor of $a$ and $b$ is 3. This means that $a = 3^k$ and $b = 3^l$ with $k > l \geq 1$ (note that $a \neq b$, since $(\Pi, f_{a,b})$ was demonstrated to be noncyclic). Then, since $(\Pi, f_{3^k, 3^l})$ is reduced, we must have $l = 1$, so we will be looking at a magma of the form $(\Pi, f_{3^k, 3})$, with $k \geq 2$. In this case, though, the only possible generator is $g = 2$. Indeed, if $g \neq 2$, it is easy to prove by induction that in the series $\{G_n\}_{n \geq 0}$, $G_0$ and subsequently all $G_{nx}$'s with $n \geq 1$ will leave 2 out; for this purpose note that the magma operation $f_{3^k, 3}(x, y) = P(3^k x + 3y)$ takes only values that are 3 or more.

B. Let $|D_a \Delta D_b| = 1$. First, consider the case in which one of $D_a$ and $D_b$ is empty. Say $D_b = \varnothing$, that is, $b = 1$. Then $a = p^k$ for some $k \geq 1$, while Proposition 3.34 evaluates the generator of $(\Pi, f_{p^k, 1})$ as $g = p$. The first item generated after $g$ will be $P(p^k g + g) = P(p^k(p + 1))$, in which case, if $P(p + 1) \leq p$, we cannot generate anything other than $g = p$. Thus it is necessary that $P(p + 1) > p$. For example, this restriction rules out cyclicity for the magmas $(\Pi, f_{p,1})$ with $p$ odd. It also rules out cyclicity for some magmas $(\Pi, f_{p^k, 1})$ with $k > 1$, e.g., $\left(\Pi, f_{19^3,1}\right)$ is not cyclic, since $P(19^3 + 1) = 7 \leq 19$.

Let us now consider the remaining possibility, in which $|D_a \Delta D_b| = 1$ with both $D_a$ and $D_b$ nonempty. Then $a$ and $b$ have 3 as a common factor, and in addition, either $a$ or $b$ has an additional prime factor $r \neq 3$, with 3 and $r$ being the only possible primes appearing in $ab$. Then the generator is $g = r$ from Proposition 3.34. We will show that necessarily $r = 2$. Indeed, if $r \neq 2$, proceeding with the ascending series $\{G_n\}_{n \geq 0}$ from the generator $g = r$, every new element produced in the process, being of the form $P(ax + by)$ with $x, y$ previously generated, is a multiple of 3, which means that we would never be able to produce the prime 2. Also, since the magma $\left(\Pi, f_{a,b}\right)$ is reduced, $a$ and $b$ cannot both be divisible by 9. Therefore, either

- $a = 3^k 2^l$ (for some $k, l \geq 1$), $b = 3$ and $g = 2$, or
- $\{a, b\} = \{3^k, 3 \cdot 2^l\}$, or $a = \max(3^k, 3 \cdot 2^l)$ and $b = \min(3^k, 3 \cdot 2^l)$ (due to the magma being reduced, we must have $a \geq b$) and $g = 2$.

Thus we have accounted for all four types listed in Theorem 3.35. Therefore if $\left(\Pi, f_{a,b}\right)$ is cyclic and reduced, it must fall in one of the four categories, thus concluding the proof.

It is now time to turn to computational matters, by selecting magmas satisfying the necessary conditions for cyclicity derived above, and calculating enough terms of $\{G_n\}_{n \geq 0}$ until the cyclic character of $\left(\Pi, f_{a,b}\right)$ becomes plausible. But of course, the beauty is in the eye of the beholder, and here comes a word of warning: building a computational case for cyclicity of a structure doesn't mean that we will have *proved* its cyclicity. In computational analyses, it is best always to keep an eye open to falsification.

Our results are detailed in four tables, each one presenting data pertaining to a structure $\left(\Pi, f_{a,b}\right)$ of one of the four types listed in Theorem 3.35.

The analyses in (Caragiu and Vicol 2016) were done in MATLAB and, in the last stages, in Julia running on Google servers. Each table lists values $g_n = |G_n|$ and $\gamma_n$, the maximum cardinality of an initial segment of primes included in $G_n$ that we could get, given our available computational resources. At Ohio Northern University, the computations were done with MATLAB R2014a on an 8 GB RAM, 64-bit Windows 7 system. For the last values of $n$ they were executed by one of the authors (P.A. Vicol) on Google servers using Google Compute Engine. The Julia program ran on machines with two virtual CPUs and 7.5 GB RAM. Running Julia in the cloud was instrumental in advancing up to the last level in each of the tables below. One of the authors (M. Caragiu) acknowledges and thanks the Getty College of Arts and Sciences at Ohio Northern University for the support provided for this project, in the form of a 2015 summer research grant.

In each of the four cases below we succeeded in running at least 10 steps, until all of the first 500,000 primes are included in $G_n$ (and thus counted by $\gamma_n$), which we considered to offer reasonable support for cyclicity (Tables 3.1, 3.2, 3.3 and 3.4).

**Table 3.1** $(\Pi, f_{27,3})$ with $g = 2$. All primes up to 45,311,969 (the 2,735,943th prime) have rank 10 or less

| $n$ | 0 | 1 | 2 | 3 | 4 | 5 | 6 | 7 | 8 | 9 | 10 |
|---|---|---|---|---|---|---|---|---|---|---|---|
| $g_n$ | 1 | 2 | 4 | 12 | 51 | 250 | 1774 | 11,765 | 74,802 | 518,620 | 3,611,134 |
| $\gamma_n$ | 1 | 1 | 1 | 1 | 23 | 130 | 879 | 7845 | 58,635 | 413,911 | 2,735,943 |

**Table 3.2** $(\Pi, f_{9,1})$ with $g = 3$

| $n$ | 0 | 1 | 2 | 3 | 4 | 5 | 6 | 7 | 8 | 9 | 10 |
|---|---|---|---|---|---|---|---|---|---|---|---|
| $g_n$ | 1 | 2 | 3 | 7 | 23 | 98 | 549 | 3277 | 21,154 | 138,373 | 885,642 |
| $\gamma_n$ | 0 | 0 | 3 | 4 | 10 | 60 | 279 | 2006 | 14,192 | 103,429 | 756,963 |

**Table 3.3** $(\Pi, f_{18,3})$ with $g = 2$

| $n$ | 0 | 1 | 2 | 3 | 4 | 5 | 6 | 7 | 8 | 9 | 10 |
|---|---|---|---|---|---|---|---|---|---|---|---|
| $g_n$ | 1 | 2 | 4 | 12 | 63 | 430 | 2662 | 15,374 | 91,649 | 557,357 | 3,448,782 |
| $\gamma_n$ | 1 | 1 | 1 | 1 | 25 | 243 | 2515 | 15,311 | 91,424 | 557,075 | 3,448,409 |

**Table 3.4** $(\Pi, f_{9,6})$ with $g = 2$

| $n$ | 0 | 1 | 2 | 3 | 4 | 5 | 6 | 7 | 8 | 9 | 10 | 11 |
|---|---|---|---|---|---|---|---|---|---|---|---|---|
| $g_n$ | 1 | 2 | 4 | 11 | 48 | 222 | 905 | 3798 | 16,296 | 71,363 | 317,015 | 1,425,352 |
| $\gamma_n$ | 1 | 1 | 3 | 6 | 20 | 204 | 853 | 3717 | 16,155 | 71,001 | 316,371 | 1,423,647 |

We present below a set of MATLAB codes representing functions that were used in the exploration of the structures $(\Pi, f_{a,b})$:

- The following function receives as input a vector $x$ and the parameters $a, b$. It calculates the extension of $x$ with the values resulting from applying $f_{a,b}$ to all possible pairs of elements of $x$. In terms of the mathematical notation previously introduced, it calculates the iterative step $X \mapsto X \cup \{P(q, r) | q, r \in X\}$.

```
function [groupoid] = groupoid(x,a,b)
N=size(x,2);
K=N;
y=x;
for I=1:N;
    for J=1:N;
        u=gpf(a*x(I)+b*x(J));
        K=K+1;
        y(K)=u;
    end
end
groupoid=sort(distinct(y));
end
```

For example, we can proceed hands-on with the calculation of $G_n$ in $(\Pi, f_{2,1})$ for $1 \leq n \leq 4$ starting from the conjectural generator 2, i.e., from $G_0 = \{2\}$:

```
>> x = 2
x =
   2
>> groupoid(x,2,1)
ans =
   2   3
>> groupoid(ans,2,1)
ans =
   2   3   7
>> groupoid(ans,2,1)
ans =
   2   3   7   11   13   17
>> groupoid(ans,2,1)
ans =
   2   3   5   7   11   13   17   19   23   29   31   37   41   43   47
```

- The following MATLAB code is based on the previous one and executes $m$ steps of the above procedure.

```
function [groupoidrec] = groupoidrec(x,a,b,m)
for I=1:m;
x=groupoid(x,a,b)
end
groupoidrec=sort(distinct(x));
```

For example, let's start with the two-elements set $X_0 = \{2, 17\}$ and use the above code to perform three steps of the iteration $X \mapsto X \cup \{P(6p + 9q) | p, q \in X\}$:

```
>> x = [2,17]
>> groupoidrec(x,6,9,3)
ans =
 Columns 1 through 16
   2    3    5    7   11   13   17   19   23   29   31   37   41   43   47   53
 Columns 17 through 32
  59   61   67   71   73   79   83   89  101  107  131  137  139  149  151  163
 Columns 33 through 48
 167  173  179  191  193  197  211  223  229  233  239  241  251  257  263  269
 Columns 49 through 52
 293  317  347  353
```

- The following code calculates the size of the greatest initial segment of primes that is included in a nonempty set of primes $X$. For example, applying this to the sets $G_n$ would provide us with the values $\gamma_n$ used in the above tables.

```
function [primesegment] = primesegment(x)
N=size(x,2);
y=primes(x(N));
z=setdiff(y,x);
if numel(z)==0
    primesegment=N;
else
w=primes(z(1)-1);
primesegment=size(w,2);
end
```

```
For example:
≫ x = [2,3,5,7,11,13,17,19,29,53]
≫ primesegment(x)
ans =
    8
```

- The next MATLAB code will produce the cardinalities $g_n$ of the sets $G_n$ recursively defined in the magma $(\Pi, f_{a,b})$ with conjectured generator "gen":

```
function [groupoidexp] = groupoidexp(gen,a,b,N)
x=[gen];
for I=1:N;
    g(I)=size(groupoidrec(x,a,b,I),2);
end;
groupoidexp=g;
```

For example, obtaining the elements $g_1, \ldots, g_8$ corresponding to the magma $(\Pi, f_{2,1})$ with conjectured generator 2 becomes straightforward:

```
≫ groupoidexp(2,2,1,8)
ans =
    2    3    6    15    33    76    187    468
```

*GPF Magmas Prologue: Random Sets of Primes and Their Expanding Coefficients in the Structures* $(\Pi, f_{a,b})$

Exploring the cyclicity of magmas $(\Pi, f_{a,b})$ raises an interesting problem of a probabilistic flavor, involving the behavior of a randomly chosen finite set of primes $X \subset \Pi$ (in terms of its size) under the action of the operation $f_{a,b}$. More precisely, we are interested in the distribution properties of the following "expanding coefficients":

$$\frac{\left|X \cup \{f_{a,b}(x,y)\,|\,x,y \in X\}\right|}{|X|}.$$

Here $X$ is a finite random set of primes. We may choose $X$, for example, to be a random set of primes of a specified cardinality that is included in a certain initial segment of the set of primes. Of course, we may adopt different models for the selection procedure of $X$.

Let's use MAPLE this time. The following MAPLE procedure is useful for exploring the magmas $(\Pi, f_{a,b})$, and it will be useful in dealing with random sets of primes. It builds on a finite set $G \subset \Pi$ and produces the expanded set obtained by iterating $n$ times the operation

$$X \mapsto X \cup \{f_{a,b}(x,y)\,|\,x,y \in X\}.$$

```
> CYCL:=proc(n::integer, a::integer, b::integer, G::set)
> P:=n->max(1,max(op(factorset(n)))); f:=(x,y)->P(a*x+b*y);
X:=G;
> for r from 1 to n do SET:=X; N:=numelems(X): K:=N^2: for
s from 1 to K do X:=X union {f(SET[1+((s-1) mod
N)],SET[1+floor((s-1)/N)])} end do: end do:
> CYCL:=X;
```

As a quick check, here is our old acquaintance $G_4$:

```
> CYCL(4,2,1,{2}); numelems(%);
```

$$\{2, 3, 5, 7, 11, 13, 17, 19, 23, 29, 31, 37, 41, 43, 47\}$$

$$15$$

Clearly, with the **CYCL** procedure, the expanding coefficient of the set of primes $G$ in the structure $(\Pi, f_{a,b})$ will be

$$\frac{\text{numelems}(\text{CYCL}(1, a, b, G))}{\text{numelems}(G)}.$$

Let's move on to random sets of primes. Possibly the first instinct would suggest, for given integers $N$ and $K$ with $0 < K < N$, a MAPLE line of code including

```
> convert([seq(ithprime(rand(1..N)()),r=1..K)],set);
```

But there is a catch! Although unlikely, especially, and increasingly if $N$ is a large integer and $K$ is fixed, a collision risk remains on the table! In addition, for large values of $N$ the function **ithprime** is rather slow.

So, we need a piece of MAPLE code that will alleviate this risk. The following MAPLE procedure **RSP**, *random set of primes*, although probabilistic, does the job of making sure that we generate a random set of $K$ primes not larger than the next prime following $N$.

```
> RSP:=proc(K::integer, N::integer)
> L:={};
> while numelems(L)<K do
> L:=convert([seq(nextprime(rand(1..N)()),r=1..K)],set);
> end do;
> RSP:=L;
> end proc;
```

Using the **nextprime** function instead of **ithprime** in **RSP** certainly adds to the speed. It quickly produces random sets of large primes in the specified range. For example:

```
> RSP(10,10^12);
```

{59655897271, 321725059451, 432323103427, 438923958457, 491548578239, 548619196921, 561491816519, 574979806777, 596883233159, 962266597267}

A variation in which the primes are selected from a prime interval that is not necessarily an initial segment is the following:

```
> RSP1:=proc(K::integer, N1::integer, N2::integer)
> L:={};
> while numelems(L)<K do
> L:=convert([seq(nextprime(rand(N1..N2)()),r=1..K)],set);
> end do;
> RSP1:=L;
> end proc;
```

With $0<K<N$ given, let us obtain (descriptive) statistics of the expanding coefficients in the magma $(\Pi, f_{a,b})$ by performing $T$ times the following computer experiment:

**[E]** Choose a random set of primes $X \subset \{p_1, p_2, \ldots, p_N\}$ with $|X| = K$, and calculate $h := |X \cup \{f_{a,b}(x,y)|x,y \in X\}|$. The output will be the "expanding coefficient" $\frac{h}{K}$.

The following procedure will provide a histogram for the $T$ expanding coefficients obtained from repeatedly applying the above experiment. Its MAPLE code uses the previous procedure **RSP**:

```
> expandingstat:=proc(N::integer, K::integer, T::integer,
a::integer, b::integer)

> for r from 1 to T do G:=RSP(K,N);
h:=numelems(CYCL(1,a,b,G)): x(r):=evalf(h/K); end do:
L:=[seq(x(r),r=1..T)];

> expandingstat:=Histogram(L, frequencyscale=absolute);

> end proc;
```

For example, selecting at random sets of 50 primes from among the first 10,000 primes and repeating this $(\Pi, f_{2,1})$ expanding experiment 200 times will give the following result.

```
> expandingstat(10^12,50,200,2,1);
```

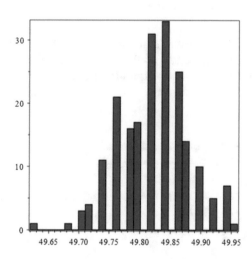

The **expandingstat** procedure can be used with different options for the histogram plot of the expanding coefficients. For example, using the "regular" MAPLE Histogram for 1000 trials with random sets of primes of cardinality 75 in $(\Pi, f_{2,1})$ selected from among the first 100,000 primes would produce the following histogram plot, apparently bimodal:

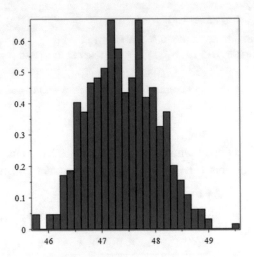

It is conceivable that in some cases it will become necessary to change the domain from which the selection of the random set of primes $K$ is made. For example, the following procedure, **expandingstat12**, is similar to **expandingstat** with the difference of having the selected random set $X$ satisfy $X \subseteq \left\{ p_{N_1}, p_{N_1}+1, \ldots, p_{N_2} \right\}$.

```
> expandingstat12:=proc(N1::integer, N2::integer, K::integer,
T::integer, a::integer, b::integer)
> for r from 1 to T do G:=RSP(K,N);
h:=numelems(CYCL(1,a,b,G)): x(r):=evalf(h/K); end do:
L:=[seq(x(r),r=1..T)];
> expandingstat12:=Histogram(L);
> end proc;
```

For example, using the above procedure in the magma structure $\left( \Pi, f_{8,3} \right)$ to conduct 1000 random selections of sets of 20 primes with ranks between 10,000 and 20,000 will produce the following output.

```
> expandingstat12(10000,20000,20,1000,8,3);
```

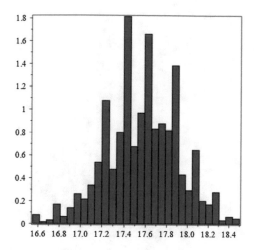

The distribution is rather narrow. The following is in the structure $(\Pi, f_{12,7})$ in which the expanding coefficient of a random set of 20 primes in $\left[p_{1,000,000}, p_{2,000,000}\right]$ appears, after 1000 experiments, to be distributed significantly narrowly and close to the upper end of 20.

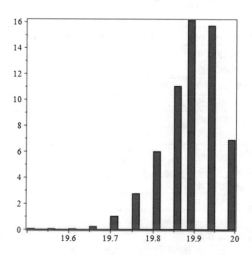

Changing the number of elements in the random set from 20 to 30, with all other parameters the same as in the previous example produces a rather similar phenomenon in which the average moves close to 30 (the cardinality of the random sets used in the experiment).

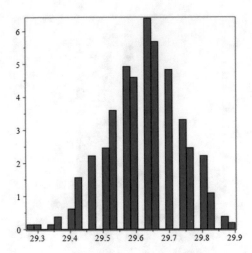

The following table presents numerical data involving the average expanding coefficient (as a result of 1000 trials) corresponding to random subsets of primes less than $10^{12}$ that have 20 elements and 30 elements, respectively, in each of the five magmas involved in our analysis of the cyclicity conjectures: $(\Pi, f_{2,1})$, $(\Pi, f_{27,3})$, $(\Pi, f_{9,1})$, $(\Pi, f_{18,3})$, $(\Pi, f_{9,6})$.

| $K$ | $a$ | $b$ | Expanding coefficient average, 1000 trials with $K$ random primes up to $10^{12}$ |
|-----|-----|-----|-----------------------------------------------------------------------------------|
| 20  | 2   | 1   | 19.98920000 |
| 20  | 27  | 3   | 19.99130000 |
| 20  | 9   | 1   | 19.99030000 |
| 20  | 18  | 3   | 19.99290000 |
| 20  | 9   | 6   | 19.99390000 |
| 30  | 2   | 1   | 29.96343334 |
| 30  | 27  | 3   | 29.97073334 |
| 30  | 9   | 1   | 29.96836668 |
| 30  | 18  | 3   | 29.97803334 |
| 30  | 9   | 6   | 29.97713334 |

This concentration of the expanding coefficient toward the size $K$ of the random set of primes appears rather intuitive (of course, intuitive does not mean proved; indeed, the level of difficulty is unknown to the author), and at least heuristically can be addressed as follows.

For a "generic" structure $(\Pi, f_{12,7})$ and a random set $X$ of a reasonably small cardinality $K$ compared to the size $N$ of the larger set of primes from which $X$ is selected, the dominant $\{f_{a,b}(x,y)|x,y \in X\}$ probably has (or intuitively is expected to have) cardinality $\approx K^2$ due to the fact that the values $f_{a,b}(x,y) = P(ax+by)$

obtained from applying the (generally) noncommutative binary operation $f_{a,b}$ to two random primes $x$, $y$ are "generically" distinct, and there are $K^2$ such pairs $(x, y)$.

**Computational Exploration Project CEP13** Gather extensive computational data and describe the asymptotic statistical behavior of the expanding coefficients

$$\frac{\left|X \cup \{f_{a,b}(x, y) | x, y \in X\}\right|}{|X|}.$$

for random sets of primes $X$ in magmas $\left(\Pi, f_{a,b}\right)$.

# Chapter 4
# Conway's Subprime Function and Related Structures with a Touch of Fibonacci Flavor

*That's a curious thing about the nature of mathematical existence. There is this abstract world which in some strange sense has existed throughout eternity. This rule hasn't physically existed in any sense in the world before a month ago, before I invented it, but it sort of intellectually existed forever.*

—John H. Conway, as cited in Siobhan Roberts's *Sciences Live: John Conway* (Roberts 2014)

While the previous chapter was primarily devoted to recurrences involving sequences of primes, or sequences of vectors with prime components, or sequences of sets of primes in the context of algebraic structures defined on the set of primes, all of them set up in terms of the greatest prime factor function, in the present chapter we will address special recurrent sequences of integers defined in terms of two other number-theoretic functions that involve in an essential way in their definitions the prime factors of their argument. These are the Euler's phi (or totient) function, and Conway's subprime function.

Everything will be placed in the inspiring context of the Fibonacci sequence, or related classical linear recurrent sequences.

## 4.1 An Euler–Fibonacci Sequence

The Euler–Fibonacci sequence was introduced at Ohio Northern University in the winter quarter of 2010–2011, in the context of a senior research project pursued in cooperation with a talented Mathematics Education major, Ashley Risch. The idea was to invent a new sequence that incorporates both Fibonacci recursion (a perennial theme in the curriculum of Mathematics Education majors) and Euler's totient function, which, even if not typically, at least potentially may be discussed in the number theory section of the "Fundamental Mathematics" sequence for prospective teachers.

© Springer International Publishing AG 2017
M. Caragiu, *Sequential Experiments with Primes*,
DOI 10.1007/978-3-319-56762-4_4

The interface between the Fibonacci sequence and Euler's phi function has been already explored. See, e.g., Luca (1999, 2000), Luca et al. (2009), Wang and Wang (2006). The "phibonacci numbers," i.e., integers $k \geq 3$ satisfying $\varphi(k) = \varphi(k-1) + \varphi(k-2)$, were introduced by A. Bager in a 1980 *American Mathematical Monthly* problem (Bager 1980), and discussed at length in Luca and Stanica (2007), where the authors prove that the sum of the inverses of all prime phibonacci numbers is finite using an appropriate sieve method, and formulate a conjecture to the effect that the sum of the inverses of all phibonacci numbers is finite. Another interesting object at the above-mentioned interface is the sequence of the values taken by Euler's phi function on Fibonacci numbers $\{\varphi(F_n)\}_{n \geq 1}$ (this goes under the OEIS listing as A065449; for references and open problems see https://oeis.org/A065449).

While the definition of phibonacci numbers in Luca and Stanica (2007) is predicative in nature, the Euler–Fibonacci sequence introduced here will be defined recursively, in an intrinsic way (not built on a recurrent sequence already in place), through a definition reminiscent of the classical Fibonacci sequence.

**Definition** The Euler–Fibonacci sequence $\{X_N\}_{n \geq 0}$ is defined as follows:

$$\begin{cases} X_0 = 0, X_1 = 1 \\ X_n = \varphi(X_{n-1} + X_{n-2} + 1) \quad \text{for} \quad n \geq 2 \end{cases} \tag{4.1}$$

The choice of (4.1) was made after we first looked into the apparently more "Fibonacci-looking" recurrence

$$x_n = \varphi(x_{n-1} + x_{n-2}) \quad \text{for} \quad n \geq 2. \tag{4.2}$$

However, (4.2) turned out to be a fairly easy integer recurrence, with every sequence of nonnegative integers satisfying (4.2) ultimately (and fairly quickly) converging to a 1-cycle of the form $\{2^k\}$ for some $k \geq 0$.

Therefore, as a first reaction, we added a 1 to the argument and we started to work on (4.1). At first we were pleasantly surprised by the seemingly steady exponential growth of the sequence, and we kept our fingers crossed in the hope that we would discover the consecutive quotients approaching some familiar constant.

Here are the terms $k \geq 0$, calculated with a simple MAPLE line:

0, 1, 1, 2, 2, 4, 6, 10, 16, 18, 24, 42, 66, 108, 120, 228, 348, 576, 720, 1296, 2016, 3312, 5256, 7200, 12456, 17860, 25200, 40256, 37368, 39600, 72900, 112500, 185400, 282204, 364800, 517600, 805392, 1133988, 1939380, 2788176, 4727556, 6819120, 11539840, 18324852, 28220080, 46471680, 70297856, 77663160, 98640672, 173595168, 256221952

While settling some elementary properties, this sequence remained a mystery throughout our work. However, the computational results were exciting, rewarding, and compensatory, in a way, for the scarcity of formal results. Our computational exploration of the sequence $\{X_n\}_{n \geq 0}$ was published in Caragiu and Risch (2011).

The next result deals with parity and a relationship with the classical Fibonacci sequence $\{F_n\}_{n\geq 0}$.

**Proposition 4.1** (Caragiu and Risch 2011) *The following are true about* $\{X_n\}_{n\geq 0}$:

- $X_n \leq F_n$ *for every* $n \geq 0$.
- $X_n \equiv 0(\mathrm{mod}\ 2)$ *for every* $n \geq 3$.

*Proof of Proposition 4.1* It is easily verified by induction that $X_n \geq 1$ for every $n > 1$. Also, $\varphi(x) \leq x - 1$ for every $x \geq 2$. Indeed, $x$ has a prime factor, and thus $\varphi(x) = x \prod_{p|x}\left(1 - \frac{1}{p}\right) \leq x - 1$. Thus for every $n \geq 2$ we have

$$X_n = \varphi(X_{n-1} + X_{n-2} + 1) \leq X_{n-1} + X_{n-2}. \tag{4.3}$$

Via a straightforward induction argument, the "sub-Fibonacci" inequality (4.3) together with the fact that $X_i = F_i$ for $i = 0, 1$ has, as a direct consequence, the inequality $X_n \leq F_n$ for every $n \geq 0$.

For the parity statement, first note that Euler's totient function takes an even value for every argument that is 3 or greater. Thus, for $n \geq 3$ we have

$$X_n = \varphi\left(\underbrace{X_{n-1} + X_{n-2} + 1}_{\geq 3}\right) \equiv 0(\mathrm{mod}\ 2).$$

This concludes the Proof of Proposition 4.1.

In the context of the Fibonacci recurrence, an interesting question that may be raised is whether integers $n$ exist with the property that three consecutive terms of the sequence (4.1) satisfy $X_n = X_{n-1} + X_{n-2}$. Indeed, such $n$'s exist, and they are not even too sparse. The ones we found, up to 500, are 2, 3, 5, 6, 7, 8, 11, 12, 13, 15, 16, 17, 19, 20, 21, 24, 31, 32, 38, 40, 84, 110, 113, 116, 124, 129, 138, 160, 178, 190, 199, 257, 287, 298, and 342.

The computational data for the Euler–Fibonacci sequence $\{X_n\}_{n\geq 0}$ was obtained with MAPLE until the amount of calculating time needed to obtain the Euler phi function involved in $X_n$ started to become too long. Then we turned for help online, continuing to calculate new terms, up to $X_{500}$ with the help of Dario Alejandro Alpern's online Elliptic curve factorization applet (DAAECF) at https://www.alpertron.com.ar/ECM.HTM.

DAAECF turned out to be an amazingly efficient and user-friendly tool for our purposes. We strongly recommend it, and we wish to thank the author for the amazing service done to number theory enthusiasts.

For example, factoring $2^{997} - 1$ with DAAECF with the derived products being the sum of the divisors, Euler's totient function, the representation as a sum of four squares (according to Lagrange's four-squares theorem, proved in 1770, to the effect that every nonnegative integer can be represented as a sum of four squares of integers), and Möbius's function, took "0 d 0 h 0 m 41.1 s" and produced the following results:

**Factorization using the Elliptic Curve Method**

133938575898283415118553131132500226320175601463191700930468798 54629388139061701531164979735
196198226594933411469414335314839316071153925544980721968373218 504918209718530288731776343325
632796392734744272769130809372947742658424845944895692993259632 864321399559710817770957553372
89565780483546507085086671 =

1675608165140848194887377679762631504050951915547329026076071 x
7993430605360222292860936960123884061988016846627213757686888797 6005930025638602973712891518592
87894468775962208410650878341385577817736702215887892074141370 00868182301410439178049533828
0826515131609456070188748300409784532283788166473583346815553


## Number of divisors: 4

Sum of divisors:
133938575898283415118553131132500226320175601463191700931268141 606829903619903109007662185758
16078395061205549046283215074898616327539955282109637154302007 95179580593937135381961048181
41097375975690315168987222307381592495983528512294522682134245193 829885925154918133876650844
53353619601935637760928832


## Euler's totient:

133938575898283415118553131132500226320175601463191700929669455 485757859161330921615637408945
76318058128661273892545455554780015814767895807851806782444429 058040613497689223643942204698
546190257125853938483893956720795603570144067668461591651768137 903439398678724542031485990125
7779413651573764092451 2

Möbius: 1

Sum of squares: a^2 + b^2 + c^2 + d^2

a =

8692055469187334276854697896923818578850144144708357768122248920611885627336790027763491295799083628998300722815181117986802386613787325957343443924330

b =

5515715489581467104401553070485261351019428691298975906629029056258133254463876033644095898992454557061823580873467145990237923284248136378216238226910

c =

5255912333169642891894720977643611267553078865752156703061492116859817795886451928593458974117309675691695844335494606251934090187200151034447585786490

d =

5822506357920811512536484044623583133960596789635030631950540637489273202222697039805822695020668151738873879528123681369852246597839538778453413295000

Returning to the Euler–Fibonacci sequence $\{X_n\}_{n \geq 0}$, we will start by pointing out an intriguing anomaly: exploring the terms $X_0, X_1, \ldots, X_{500}$ showed that the pattern $X_n < X_{n+1}$ holds almost everywhere, though with the following exceptions:

$$40256 = X_{27} > X_{28} = 37368,$$

$$3085481349077224029931607704 = X_{149} > X_{150}$$
$$= 3072057034451510523677529600,$$

$$62158007546211147073048654711600 = X_{180} > X_{181}$$
$$= 56847317174037843272516890252800,$$

$$56877065807749322466261174046848000 = X_{192} > X_{193}$$
$$= 56013460186771626017960239968096000,$$

## 4.1.1    A Kepler Moment?

In 1608, in a letter to Joachim Tankius, Johannes Kepler communicated his observation that the sequence of consecutive Fibonacci quotients $\frac{F_{n+1}}{F_n}$ approaches the golden section (Curchin and Herz-Fischler 1985). Since it is inspired by the classical Fibonacci sequence, here is a question that should be asked about the Euler–Fibonacci sequence $\{X_n\}_{n \geq 0}$, namely, *Do we have a Kepler moment here?*

In other words, if we define, for $n \geq 1$, $Q_n = X_{n+1}/X_n$ what can be said about $\lim_{n \to \infty} Q_n$?

The data acquired with MAPLE and the DAAECF applet allowed us to visualize a plot of the quotients $X_{n+1}/X_n$ for $n < 500$:

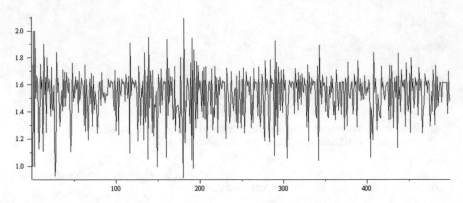

There may be something here, but we couldn't sense a definitively clean-cut "eureka." Yes, the consecutive quotients evolve around the golden section, $\alpha = \frac{1+\sqrt{5}}{2}$, and there is hope that with more computational data, the oscillations would be seen more visibly converging to $\alpha = 1.6180339\ldots$.

A little more encouraging is the histogram of the set of consecutive quotients of the Euler–Fibonacci sequence. However, the skewness of this particular distribution brings down its average to a value (1.51448) smaller than $\alpha$.

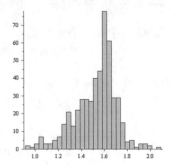

This can also be seen in the plot of the sequence $\frac{\ln X_n}{n}$ for $1 \le n \le 500$, displayed below together with the graph of the constant function $\ln \alpha = 0.4812118246\ldots$.

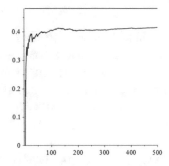

Next we will discuss an interesting congruential property of $\{X_n\}_{n \ge 0}$.

## 4.1.2 The Euler–Fibonacci Sequence Modulo 4

Another interesting observation is that the congruence $X_n \equiv 0 \pmod 4$ holds for $13 \le n \le 500$. We may ask whether hypothetically there exists $n > 500$ such that the $n$th term of the sequence satisfies $X_n \equiv 2 \pmod 4$. What is the likelihood of this? Let us denote by $m$ the least integer $m > 500$ such that $X_m = \varphi(X_{m-1} + X_{m-2} + 1) \equiv 2 \pmod 4$. Then the quantity $A := X_{m-1} + X_{m-2} + 1$ must have the following properties:

- The sum $A$ must be congruent to 1 modulo 4 (from the minimal property of $m$, both terms $X_{m-1}$ and $X_{m-2}$ must be multiples of 4).
- The sum $A$ must not have any prime divisor congruent to 1 modulo 4; otherwise, $\varphi(A) \equiv 0 \pmod 4$.
- Finally, the sum $A$ must be of the form $A = p^{2k}$, where $p$ is a prime congruent to 3 modulo 4. Indeed, $A$ cannot have two prime divisors congruent to 3 modulo 4, since otherwise, $\varphi(A) \equiv 0 \pmod 4$, and if $p \equiv 3 \pmod 4$ is the necessarily unique prime factor of $A$, it must appear to an even power, since $A \equiv 1 \pmod 4$.

So the first $m > 500$ with $X_n \equiv 2 \pmod 4$ appears when $X_{m-1} + X_{m-2} + 1 = p^{2k}$ for some prime $p \equiv 3 \pmod 4$.

For us, the Euler–Fibonacci sequence remains, at least for now, a mystery with "potential," with practically all nontrivial questions about it remaining open. Are there infinitely many "drops" $X_{n+1} < X_n$? Are there infinitely many solutions to the equality $X_n = X_{n-1} + X_{n-2}$? Is it really true that $\lim_{n \to \infty} Q_n = (1 + \sqrt 5)/2$? If not, what can be said about $\lim_{n \to \infty} Q_n$, or if that does not exist, what about the corresponding $\liminf_{n \to \infty} Q_n$ and $\limsup_{n \to \infty} Q_n$? Is there any $n > 500$ with $X_n \equiv 2 \pmod 4$? Perhaps one way to move forward lies in obtaining more data through more intense computational work. This brings us to the next CEP:

**Computational Exploration Project CEP14** Shed more light into the above questions about the sequence (4.1) by calculating more terms $X_n$.

In order to help those who intend to continue with the computational approach, we refer to a link where the terms $\{X_n\}_{n=1}^{500}$ are stored:

http://mihai-caragiu-maths.blogspot.com/2010/10/phi-bonacci-sequence-update.html

The last two terms in the sequence (enough to restart the computation) are the following:

$X_{499} = 9170869910159854433983456646110267914795097481611303724018954690701365290480081561586201 60$

$X_{500} = 1357870193719059716256430397688783608072366877233385070201855578490811022106408391147520000$

## 4.2  Conway's Subprime Fibonacci Sequences

Borrowing John Conway's own formulation, the "subprime Fibs" (subprime Fibonacci sequences) are some truly amazing "*wild beasts*" of the experimental mathematics realm. It appears that they were introduced by John Conway (Roberts 2014; The Mathematical Tourist blog 2013), the celebrated mathematical genius and emeritus professor at Princeton while he was on a plane on his way to visit Guy et al. (2014). These second-order recurrent sequences are defined as follows:

$$\begin{cases} x_0, x_1 \in \mathbb{N} \text{ (initial conditions)} \\ x_n = C(x_{n-1} + x_{n-2}) \quad \text{for} \quad n \geq 2 \end{cases} \tag{4.4}$$

The function $C(x)$ is defined in a very simple way: if $x$ is 1 or a prime, leave it as it is, otherwise just divide $x$ by its least prime factor.

The class sequences defined above have a remarkable property: no matter how we choose the initial conditions $x_0, x_1$ in (4.4), the calculation always leads us into a cycle. This remains a conjecture. In the words of Conway himself [as cited in Roberts (2014)],

> It seems no matter which pair of numbers you start with, it ends up being periodic, it repeats after a certain number of steps, as is the case with the sequence I just did. I can't say it always repeats. Because I haven't proved it. But it's as obvious as hell that it always does.

Let's use MAPLE to generate a visual of the terms $x_0, \ldots, x_{100}$ of (4.4) with $x_0 = x_1 = 1$:

```
> C:=proc(n::integer) local u: if isprime(n)='true' or n=1
then n else u:=factorset(n): n/min(seq(u[j], j=1..nops(u)))
end if end:

> x(0):=0; x(1):=1; for r from 2 to 100 do x(r):=C(x(r-1)+x(r-
2)) end do: L:=[seq(x(r),r=0..100)]: listplot(L);
```

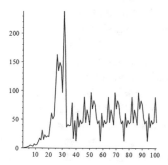

I first found out about the subprime Fibs, specifically about a terrific paper dealing with them in the summer of 2012, after a brief and enriching email exchange with Richard Guy. At the time, the paper was available on arXiv.org. It was subsequently published, with a dedication to Martin Gardner, in *Mathematics Magazine* (Guy et al. 2014) in 2014. Six nontrivial cycles for subprime Fibonacci sequences are discovered in Guy et al. (2014) for the initial conditions $1 \leq x_0, x_1 \leq 100,000$, their lengths being

$$10, 11, 18, 19, 56, 136$$

In Guy et al. (2014), the authors also present, in a standard display of experimental mathematics virtuosity, an illuminating table featuring the distribution of cycle lengths in terms of the number of pairs of initial conditions leading to a limit cycle of a given length, to the effect that in more than 94% of the cases, the limit cycle has size 18 or 136 (slightly more for 136 than for 18). An interesting informal argument that supports the ultimate periodicity of Conway's subprime sequences (which remains yet a conjecture) is also provided in Guy et al. (2014).

Whether there are limit cycles other than those corresponding to the six lengths listed above is not known.

I was amazed and encouraged by the convergence of ideas (in itself, a sort of mathematical universality) with the ultimate periodicity conjecture for greatest prime factor sequences, discussed here earlier in Chapter 3. For instance, in Back and Caragiu (2010) it was proved that all GPF-Fibonacci sequences (prime sequences in which every term is the greatest prime factor of the sum of the two preceding terms) starting from a pair of distinct primes $x_0 \neq x_1$ must enter the limit cycle $[7, 3, 5, 2]$.

However, since the subprime function is most of the time strictly larger than the greatest prime factor function, it is natural to expect that a proof of the ultimate periodicity of subprime Fibs will be harder than the proof of the ultimate periodicity of the GPF-Fibonacci sequences. Moreover, in general, we expect problems involving Conway's subprime function to be harder than their greatest prime factor counterparts.

Another result along similar lines appeared in a 2016 *Fibonacci Quarterly* paper, "Prime Fibonacci Sequences" (Alm and Herald 2016). The authors use a definition modeled after subprime Fibs, in which every term is the smallest odd prime divisor of the sum of the two preceding terms (also, the sequence is set to terminate if it gets to two terms that sum to a power of 2). The authors prove that every such "Prime Fibonacci Sequence" terminates in a power of 2, and they prove a series of interesting related results, including the possibility of extending every such sequence indefinitely to the left. What I find remarkable about this article is that it is a nice result of faculty–student research in number theory at a small undergraduate college (Illinois College, a private liberal arts institution in Jacksonville, Illinois), with one of the coauthors a senior, on the dean's list in June 2016.

### 4.2.1  A Monte Carlo Approach to Subprime Fib Period Search

We will now apply Floyd's tortoise and hare cycle-finding algorithm to verify or falsify the hypothesis of the existence of seven possible limit cycles for Conway's subprime Fibonacci sequences (including the trivial type) 1, 10, 11, 18, 19, 56, and 136. Note that this will not detect without further refinement the existence of disjoint nontrivial cycles of the same length.

We first set up the Conway's subprime iteration on vectors:

```
> f:=(x,y)->(y,C(x+y));
```

We now use Floyd's algorithm for $M$ trials with random initial conditions picked from 1 to $T$. For each trial we record the cycle length. This will produce a vector PER with $M$ elements. We find the number of occurrences of 1, 10, 11, 18, 19, 56, and 136 in PER, and to have a summary view, we convert PER to a set to eliminate duplicates.

Below we show the algorithm performing $M = 100{,}000$ trials with initial conditions randomly chosen from 1 to $10^{12}$. In this Monte Carlo experiment, the clause

$$\text{is } (P1 + P10 + P11 + P18 + P19 + P56 + P136)$$

produces *true* as output and thus validates the hypothesis that the items 1, 10, 11, 18, 19, 56, and 136 cover all possibilities for the size of a limit cycle of a subprime Fib sequence.

Repeated trials with various fairly large values of $M$ and $T$ showed that cycles of lengths 1 and 10 become increasingly rare.

At the same time, they confirmed the trend of about 47% for the proportion of initial conditions generating subprime Fibonacci sequences eventually entering each of the limit cycles of length 18 and 136, included in the computational results listed in Table 1 of Guy et al. (2014).

```
> M:=100000: T:=10^12: for r from 1 to M do A:=rand(1..T)();
B:=rand(1..T)(); X:=(A,B); SLOW:=f(X): FAST:=f(f(X)): k:=1:
while SLOW<>FAST do SLOW:=f(SLOW): FAST:=f(f(FAST)): k:=k+1: end
do: TENTATIVE_START:=k: X:=SLOW: SLOW:=f(X): FAST:=f(f(X)):
k:=1: while SLOW<>FAST do SLOW:=f(SLOW): FAST:=f(f(FAST)):
k:=k+1: end do: PERIOD[r]:=k: end do:
PER:=[seq(PERIOD[r],r=1..M)]: P1:=Occurrences(1,PER);
P10:=Occurrences(10,PER); P11:=Occurrences(11,PER);
P18:=Occurrences(18,PER); P19:=Occurrences(19,PER);
P56:=Occurrences(56,PER); P136:=Occurrences(136,PER);
convert(PER, set); is(P1+P10+P11+P18+P19+P56+P136=M);
```

$$P1 := 1$$
$$P10 := 8$$
$$P11 := 29$$
$$P18 := 47169$$
$$P19 := 3099$$
$$P56 := 2320$$
$$P136 := 47374$$
$$\{1, 10, 11, 18, 19, 56, 136\}$$
$$true$$

### 4.2.2 General Second-Order Subprime Sequences

A class of recurrences that seems to be "next in line" after Conway's subprime Fibonacci sequences is that of the general second-order "linear-looking" recurrences based on Conway's subprime function $C(x)$, that is, sequences satisfying

$$\begin{cases} x_0, x_1 \in \mathbb{N} \\ x_n = C(ax_{n-1} + bx_{n-2}) & \text{for} \quad n \geq 2 \end{cases}.$$

All experiments we conducted with sequences of the above type for various parameters $a$, $b$ and various initial conditions $x_0, x_1$ showed that ultimate periodicity still holds. Still, we faced a significant obstacle in acquiring data, related to the possibility of large cycle lengths emerging in some cases.

Just to give an example, while preparing my 2016 MathFest talk, I experimented with various values of $a, b, x_0, x_1$. Everything went fine. Large periods appeared and were "visible to the naked eye" in the sense that calculating a couple of thousand terms or tens of thousands of terms of $\{x_n\}_{n \geq 0}$ and plotting the list thus obtained featured immediately visual evidence of the entrance into a limit cycle.

Everything went fine until I tried, off the top of my head, $a = 2$, $b = 7, x_0 = 1, x_1 = 100$, which led to the second-order "Conway subprime" sequence

$$\begin{cases} x_0 = 1, x_1 = 100 \\ x_n = C(2x_{n-1} + 7x_{n-2}) \quad \text{for} \quad n \geq 2 \end{cases}$$

Calculating terms and looking at the listplot did not produce anything visible in the range of tens of thousands of terms. After "doubling down" (being certain that I would be able to see some sign of a cycle pretty soon) and trying to visualize longer lists of terms (using hundreds of thousands of terms, pushing MAPLE to the limit), there was still no visible sign of periodicity, and eventually the process just froze.

Clearly, the low-tech method (but rewarding to the human eye) could not claim any positive results in this case.

This called for the Floyd's period-detection algorithm. That approach finally produced validation of periodicity, with MAPLE spending a one-nighter (and beyond) to find the period, which turned out to be greater than 3 million:

$$T = 3,132,779.$$

This explained, after the fact, the lack of success in the naked-eye approach of "calculate and listplot." At the same time, it reinforces the conjecture that all such second-order Conway subprime sequences are ultimately periodic.

## 4.3   Subprime Tribonacci Sequences and Beyond

Moving on to third-order Conway subprime recurrent sequences, the first candidate will be the analogue of the classical tribonacci sequence. Recall that in Section 3.2 we explored the greatest prime factor analogue and found four nontrivial cycles for the GPF-tribonacci recurrences, with lengths 100, 212, 28, and 6. We would like to pursue a similar approach based on Conway's subprime function.

Let us first look into the "canonical" Conway tribonacci sequence, which we define as follows:

$$\begin{cases} x_n = C(x_{n-1} + x_{n-2} + x_{n-3}), \\ x_0 = 0, x_1 = 0, x_2 = 1. \end{cases}$$

For this particular third-order recurrence, we will use our standard MAPLE code (see Section 2.4) in order to derive the trio "tentative start + period + preperiod":

```
>      f:=(x1,x2,x3)->(x2,x3,C(x3+x2+x1)):      SEED:=(0,0,1):      X:=SEED:
SLOW:=f(X):   FAST:=f(f(X)):   k:=1:  while  SLOW<>FAST do  SLOW:=f(SLOW):
FAST:=f(f(FAST)):    k:=k+1;    end    do:    TENTATIVE_START:=k;    Y:=SLOW:
X:=SLOW:    SLOW:=f(X):    FAST:=f(f(X)):    k:=1:    while    SLOW<>FAST    do
SLOW:=f(SLOW):    FAST:=f(f(FAST)):    k:=k+1;    end do:  PERIOD:=k;  m:=0:
U:=SEED:    V:=Y:    while    U<>V    do    m:=m+1;    U:=f(U);    V:=f(V)    end    do:
PREPERIOD:=m;
```

Thus, we obtain at once the following:

$$\text{TENTATIVE\_START}:=3174$$

$$\text{PERIOD}:=3174$$

$$\text{PREPERIOD}:=1214$$

Thus, for every integer $k \geq 1214$ the equality $x(k+3174) = x(k)$ holds. We now have available the complete periodicity profile of the canonical Conway tribonacci sequence: it is an integer sequence with period 3174, preperiod 1214, with the maximum element in the limit cycle being 454507.

For a possible variation, one may consider a couple of other Conway subprime third-order recurrences (where $a, b, c$ are positive integers) based on the same seed:

$$\begin{cases} x_n = C(\alpha x_{n-1} + \beta x_{n-2} + \gamma x_{n-3}), \\ x_0 = 0, x_1 = 0, x_2 = 1. \end{cases}$$

Here is a MAPLE procedure, **PERQUEST**, for the calculation (in principle, since computing time is a different story) of the periods $T(\alpha, \beta, \gamma)$ of the above sequences for every possible triple $(a, b, c)$ of positive integers (for example, we just found out that $T(1, 1, 1) = 3174$):

```
PERQUEST:=proc(a::integer, b::integer, c::integer)
f:=(x1,x2,x3)->(x2,x3,C(a*x3+b*x2+c*x1)):      SEED:=(0,0,1):      X:=SEED:
SLOW:=f(X):  FAST:=f(f(X)):  k:=1:  while  SLOW<>FAST  do  SLOW:=f(SLOW):
FAST:=f(f(FAST)):    k:=k+1;    end    do:    TENTATIVE_START:=k:    Y:=SLOW:
X:=SLOW:    SLOW:=f(X):    FAST:=f(f(X)):    k:=1:    while    SLOW<>FAST    do
SLOW:=f(SLOW):  FAST:=f(f(FAST)):  k:=k+1;  end do:  PERIOD:=k;
PERQUEST:=PERIOD;

end proc;
```

On the other hand, if we let the initial conditions float, we are then considering generalized Conway tribonacci sequences, defined as follows:

$$\begin{cases} x_0 = a, x_1 = b, x_2 = c \\ x_n = C(x_{n-1} + x_{n-2} + x_{n-3}) & \text{for} \quad n \geq 3 \end{cases}.$$

Applying our Monte Carlo randomized search for possible periods of generalized Conway tribonacci sequences yielded the following periods:

- A period of 3174, as described above. By far, most random selections of initial conditions $(a, b, c)$ lead to such a limit cycle.
- A period of 6, with limit cycle consisting of (in order) 37, 139, 73, 83, 59, and 43.
- A period of 5, with limit cycle consisting of (in order) 25, 97, 53, 35, and 37.
- A period of 203. We witnessed such a period twice in a set of 1000 trials with random initial conditions $1 \leq a, b, c \leq 10^{12}$. Finding it will constitute our next computational project:

**Computational Exploration Project CEP15** Find the components of a 203-cycle pertaining to some generalized Conway tribonacci sequence.

### 4.3.1   What Lies Beyond?

Of course, much more remains to be explored when it comes to most general "linear" Conway subprime recurrences of the form

$$x_n = C(c_1 x_{n-1} + c_2 x_{n-2} + \cdots + c_d x_{n-d} + c_0) \quad \text{for} \quad n \geq d.$$

Regarding these general sequences, obviously the most important question is this:

### 4.3.2   Are All of Them Ultimately Periodic?

In that respect, one can envision two possible directions to follow at an elementary level (that is, leaving aside the possibility of proving the ultimate periodicity, through advanced analytic number theory arguments that are beyond the scope of this undergraduate number theory lab manual):

- Either we will discover or invent a surprising ansatz on the initial conditions $x_0, x_1, \ldots, x_{d-1}$ that will require that such a sequence have infinite limit (and hence no limit cycle), or

- We will gather extensive computational evidence supporting ultimate periodicity for higher-order Conway subprime recurrences of the above type, for a rich variety of orders $d$ and randomly chosen initial conditions.

Both of these directions are excellent topics for future work, but we will not explore them in this book.

However, we will not yet leave the amazing field of recurrences based on Conway's subprime function. Instead, we will address Conway subprime sequences of integer vectors, trying to replicate experimentally the result derived in Section 3.6 on GPF Ducci games.

## 4.4 Conway Subprime Ducci Games

We already know from Section 3.6 that the iteration $G : \Pi^d \to \Pi^d$ defined in terms of the greatest prime factor function $P$,

$$G(x_1, x_2, x_3, \ldots, x_{d-1}, x_d) = (P(x_1 + x_2), P(x_2 + x_{23}), P(x_3 + x_4), \ldots, P(x_{d-1} + x_d), P(x_d + x_1)),$$

always leads to a limit cycle, which if not trivial, has the property (Theorem 3.26) that the union of all vectors within the cycle (converted into sets) is precisely the set $\{2, 3, 5, 7\}$.

Designing the Conway subprime analogue of the GPF Ducci games is quite straightforward. We will be looking at the iteration $\Gamma : \mathbb{N}^d \to \mathbb{N}^d$ defined as follows (where $C$ is the Conway's subprime function):

$$\Gamma(x_1, x_2, x_3, \ldots, x_{d-1}, x_d) = (C(x_1 + x_2), C(x_2 + x_{23}), C(x_3 + x_4), \ldots, C(x_{d-1} + x_d), C(x_d + x_1)).$$

Since in dimension $d = 2$ the Conway Ducci games immediately lead to a cycle of length 1, from now on we may assume $d \geq 3$.

Here is the general MAPLE procedure that executes one iteration of the Conway Ducci game after receiving an input vector (list) $x \in \mathbb{N}^d$:

```
> ConwayDucci:= proc(x::list)::list;
> local d, k, X;
> d:=numelems(x);
> for k from 1 to d-1 do X[k]:=C(x[k]+x[k+1]) end do;
> X[d]:=C(x[d]+x[1]);
> [seq(X[k],k=1..d)];
> end proc;
```

This calculates fairly quickly:

```
> L:=[1,2,3,4,5,6,7,8,9,10]:
> ConwayDucci(L);
```

$$[3, 5, 7, 3, 11, 13, 5, 17, 19, 11]$$

Let us set up the visualization of a set of $m$ Conway Ducci iterations as an $m \times d$ array, with each new vector in the iteration placed immediately below the vector it derives from, with the "seed" vector at the top.

For an easier visualization we will equip each vector with an additional component, placed first on the left, that indicates the vector rank in the iteration.

Say we want to play a 5-dimensional Conway Ducci game, starting with the seed list

$$[27, 96, 17, 90, 34].$$

Since the rank of the seed is 0, we will assort this, in the MAPLE calculation, with a zero to the left.

```
> a(1,1):=0; a(1,2):=27; a(1,3):=96; a(1,4):=17;
  a(1,5):=90; a(1,6):=34;
```

Let's say we want to see all vectors up to rank 20 appearing in the Conway Ducci iteration starting from the above seed.

```
> for k from 2 by 1 to 20 do a(k,1):=k-1: a(k,2):=C(a(k-
  1,2)+a(k-1,3)): a(k,3):=C(a(k-1,3)+a(k-1,4)): a(k,4):=C(a(k-
  1,4)+a(k-1,5)): a(k,5):=C(a(k-1,5)+a(k-1,6)): a(k,6):=C(a(k-
  1,6)+a(k-1,2)): end do:
```

Now let us adopt the strategy of visualizing an array by associating with it a function of two variables while setting the table size to infinity (in practice, we do many more than 20 iterations):

```
> interface(rtablesize=infinity);
```

(tip: you may want to click "enter" twice on the above setting)

```
> g:=(i,j)->a(i,j); Matrix(21,6,g);
```

And then ... *voilà*:

```
> Matrix(21,6,g);
```

[[0,27,96,17,90,34],[1,41,113,107,62,61],[2,77,110,13,41,51],[3,17,41,27,46,64],
[4,29,34,73,55,27],[5,21,107,64,41,28],[6,64,57,35,23,7],[7,11,46,29,15,71],[8,19,25,22,43,41],
[9,22,47,13,42,30],[10,23,30,11,36,26],[11,53,41,47,31,7],[12,47,44,39,19,30],
[13,13,83,29,7,11],[14,48,56,18,9,12],[15,52,37,9,7,30],[16,89,23,8,37,41],[17,56,31,15,39,65],
[18,29,23,27,52,11],[19,26,25,79,21,20],[20,17,52,50,41,23]]

Visualization always gives the feeling of immediate control of the problem. However, if the problem of periodicity is to be addressed, and if the period of the Conway Ducci game is not small, then we will have a problem. Can you find the period of the game with the seed vector selected above by possibly performing a sufficiently large number of iterations?

Even if you are willing to build such an array (in which case you have to brace yourself for an array with a row number in the thousands, if not more), it is obvious that it is more convenient to use a rigorous cycle-finding method such as Floyd's algorithm written for this type of problem.

Here is the MAPLE code to produce the Conway Ducci period for this particularly chosen seed vector and the tentative starting point of the periodicity zone:

```
> SEED:=[27,96,17,90,34]: SLOW:=ConwayDucci(SEED):
FAST:=ConwayDucci(ConwayDucci(L)): k:=1: while SLOW<>FAST do
SLOW:=ConwayDucci(SLOW): FAST:=ConwayDucci(ConwayDucci(FAST)):
k:=k+1; end do: TENTATIVE_START:=k; x:=SLOW:
SLOW:=ConwayDucci(x): FAST:=ConwayDucci(ConwayDucci(x)): k:=1:
while SLOW<>FAST do SLOW:=ConwayDucci(SLOW):
FAST:=ConwayDucci(ConwayDucci(FAST)): k:=k+1; end do:
PERIOD:=k;
```

$$\text{TENTATIVE\_START}:=6040$$

$$\text{PERIOD}:=3020$$

Now we know that this particular 5-number Conway Ducci game has period 3020. What about finding the periods for all such games? Unfortunately, we don't have, as of now, a theoretical tool that would reduce the Conway Ducci iteration in a cycle to a simple format so that the possible cycle lengths could be completely determined. This was the case with the classical Ducci games (where the iteration format within the cycle was essentially binary) and with the GPF Ducci games as well (where the components of the vectors in a cycle were clearly determined to be among the numbers 2, 3, 5, 7).

The next best thing would be to use again our Monte Carlo–type technique. We present below such an exploratory search for periods of *d*-number Conway Ducci games for $3 \leq d \leq 6$, where the 1000 such games that were analyzed had the components of the seed randomly chosen up to a billion, while the period for each such game was determined by running the Floyd period-finding algorithm for the Conway Ducci iteration. For every such trial the output will be twofold:

- The set of all periods found, **CYCLE_LENGTHS**
- The sequence **OCCURRENCES** specifies the number of occurrences of each item in the set of periods among the 1000 trials executed.

The MAPLE code below illustrates the case $d = 5$ and in theory can be adapted to any dimension:

```
M:=1000: T:=10^9: for r from 1 to M do A1:=rand(1..T)();
A2:=rand(1..T)(); A3:=rand(1..T)(); A4:=rand(1..T)();
A5:=rand(1..T)(); X:=[A1,A2,A3,A4,A5]; SLOW:=ConwayDucci(X):
FAST:=ConwayDucci(ConwayDucci(X)): k:=1: while SLOW<>FAST do
SLOW:=ConwayDucci(SLOW): FAST:=ConwayDucci(ConwayDucci(FAST)):
k:=k+1: end do: TENTATIVE_START:=k; X:=SLOW:
SLOW:=ConwayDucci(X): FAST:=ConwayDucci(ConwayDucci(X)): k:=1:
while SLOW<>FAST do SLOW:=ConwayDucci(SLOW):
FAST:=ConwayDucci(ConwayDucci(FAST)): k:=k+1: end do:
PERIOD[r]:=k; end do: PER:=[seq(PERIOD[r],r=1..M)];
CYCLE_LENGTHS:=convert(PER,set);
OCCURRENCES:=seq(Occurrences(CYCLE_LENGTHS[t],PER),t=1..numele
ms(CYCLE_LENGTHS));
```

Note that a cycle of length 1 is always possible (for example, in the rare cases in which the seed of the Conway Ducci game has equal components). So, even if the cycle length 1 did not appear among the elements of the **CYCLE_LENGTHS** set, we will still enter the number of occurrences of a cycle length 1 with 0%. The fact that it did not appear in such a randomized search is understandable.

This must be recorded as one of the limitations of this Monte Carlo technique. To force the recording of the presence of a 1-cycle, one should either increase the number of trials (here 1000) in such a search, or just repeat the trial.

And now we present the results. For the sake of authenticity (the golden rule of lab work) we performed one set of 1000 random trials for each dimension. We urge the reader to try this. It is likely that new cycle lengths are waiting to be discovered! This sort of "lab work" was one of our most rewarding activities.

For $d = 3$ we obtained the following:

- Cycle length 1 in 0.0% of cases
- Cycle length 18 in 38.5% of cases
- Cycle length 33 in 0.1% of cases
- Cycle length 53 in 3.5% of cases
- Cycle length 57 in 2.7% of cases
- Cycle length 159 in 3.9% of cases

- Cycle length 168 in 1.2% of cases
- Cycle length 327 in 4.8% of cases
- Cycle length 408 in 43.0% of cases
- Cycle length 936 in 2.3% of cases

For $d = 4$ we obtained the following:

- Cycle length 1 in 1.7% of cases
- Cycle length 16 in 42.8% of cases
- Cycle length 24 in 0.1% of cases
- Cycle length 200 in 1.1% of cases
- Cycle length 328 in 22.3% of cases
- Cycle length 336 in 32.0% of cases

For $d = 5$ we obtained the following:

- Cycle length 1, found in 0.0% of the cases
- Cycle length 35, found in 0.1% of the cases
- Cycle length 95, found in 2.3% of the cases
- Cycle length 2083, found in 1% of the cases
- Cycle length 3020, found in 46.6% of the cases
- Cycle length 3845, found in 49.2% of the cases
- Cycle length 5470, found in 0.1% of the cases
- Cycle length 10415, found in 0.7% of the cases

For $d = 6$ we obtained the following:

- Cycle length 1 in 4.4% of cases
- Cycle length 12 in 1.1% of cases
- Cycle length 18 in 70.0% of cases
- Cycle length 53 in 0.6% of cases
- Cycle length 57 in 1.1% of cases
- Cycle length 78 in 0.6% of cases
- Cycle length 159 in 0.2% of cases
- Cycle length 168 in 0.5% of cases
- Cycle length 327 in 0.4% of cases
- Cycle length 408 in 17.7% of cases
- Cycle length 456 in 3.3% of cases
- Cycle length 984 in 0.1% of cases

For $d = 7$ the result of the computation revealed a substantial spike when it comes to the length of the largest period, 4347602. The results are as follows:

- Cycle length 1 in 0.0% of cases
- Cycle length 70 in 1.6% of cases
- Cycle length 1848 in 0.2% of cases
- Cycle length 7343 in 83.7% of cases
- Cycle length 16051 in 13.8% of cases
- Cycle length 4347602 in 0.7% of cases.

Finally, for $d = 8$, the Conway Ducci computation finalized after a bit more than 7 h of MAPLE continuous run and provided the following results:

- Cycle length 1 in 10.9% of cases
- Cycle length 16 in 2.1% of cases
- Cycle length 328 in 1.5% of cases
- Cycle length 336 in 3.5% of cases
- Cycle length 352 in 1.7% of cases
- Cycle length 57232 in 80.3% of cases

In the end, as a brief comparison summary between the GPF Ducci games (Section 3.6) and the Conway Ducci games, we could say the following:

➢ Theoretically speaking, we managed to prove the ultimate periodicity of the GPF Ducci games, but not for the Conway Ducci games, which is, we believe, a significantly harder problem. As mentioned previously, problems involving the Conway subprime function, especially in matters of ultimate periodicity, are harder than their "greatest prime factor" counterparts.
➢ The cycle lengths found for the Conway Ducci games reveal a higher complexity than the cycle length statistics for the GPF Ducci games.
➢ For the GPF Ducci games, the description of the periods we obtained in dimensions ranging from 3 to 8 is final. On the other hand, the nature of the Monte Carlo type of analysis used for the Conway Ducci games is a continuing project: even in the dimensions addressed so far (still ranging from 3 to 8), it is certainly possible that more powerful computations could add new cycle lengths to our listings.

**Computational Exploration Project CEP16** Further explore the periods of the Conway Ducci games in dimensions ranging from 3 to 8 (and also beyond that) by increasing the range of the components of the seed vectors and/or using alternative random number generators for those components, by increasing the number of trials in an experiment, or by working on more powerful platforms.

## 4.5   Conway Subprime Magmas, a Remarkable Cyclicity Result, and an Unexpected Sighting of the Golden Ratio

In this section we will consider a shift in the use of Conway's subprime function $C(x)$ from a paradigm involving sequences to a paradigm involving algebraic structures. To this end, we will introduce a binary operation $\oplus$ on the set of positive integers $\mathbb{N} = \{1, 2, 3, \ldots\}$ that associates to every pair $(x, y) \in \mathbb{N} \times \mathbb{N}$ the element

$$x \oplus y := C(x+y).$$

Let us agree to call this operation *Conway addition*. Conway addition displays no obvious "textbook" properties: it is not commutative, not associative, and has no identity element. This makes $(\mathbb{N}, \oplus)$ just a seemingly amorphous magma.

And yet, a closer look unveils an interesting potential. The interesting thing is that while, as previously mentioned, the use of Conway's subprime function makes recurrent sequences of the form $x_n = C(c_1 x_{n-1} + c_2 x_{n-2} + \cdots + c_d x_{n-d} + c_0)$ harder to explore (at least computationally) than their greatest prime factor counterparts $x_n = P(c_1 x_{n-1} + c_2 x_{n-2} + \cdots + c_d x_{n-d} + c_0)$, and while the same goes for the related Ducci games, in that the Conway Ducci games $X_k = \Gamma(X_{k-1})$ are harder to explore than their greatest prime factor counterparts $X_k = G(X_{k-1})$, for which we actually found a general characterization of the limit cycles, when it comes to the corresponding algebraic structures, we will discover an interesting reversal.

Recall that proving cyclicity for the special greatest prime factor magmas $(\Pi, f_{a,b})$ was a hard problem. While we know that in order to be cyclic, a GPF magma with the operation $f_{a,b}(x, y) = P(ax + by)$ on the set of primes needs to satisfy such and such conditions in order to stand a chance of cyclicity (see the necessary conditions listed in Theorem 3.35), we were not able to formally prove any cyclicity result involving a particular structure $(\Pi, f_{a,b})$.

The next theorem will show that if we replace the greatest prime factor function with Conway's subprime function, we do find a "Conway subprime magma" for which we can actually prove a cyclicity result (Caragiu et al., to appear in Fibonacci Q.).

**Theorem 4.2** *The structure* $(\mathbb{N}, \oplus)$ *is cyclic, generated by the element 1, i.e.,*

$$\mathbb{N} = \langle 1 \rangle.$$

*Let us begin with a computational exploration, just to get a sense of the manner in which a single element, 1, generates every positive integer via the operation* $\oplus$.

MATLAB will be a good choice. In this particular context, the least prime factor function is essential. Let us recall its code:

```
function lf=lf(n)
lf=min(factor(n));
end
```

Once this is up, we will able to calculate Conway's subprime function with the following code:

```
function conway=conway(n)
if n==1
conway=1;
elseif isprime(n)==1;
    conway=n;
else
    conway=n/lf(n);
end
```

The next MATLAB code receives a set $X$ as input and calculates the basic iterative step in the magma generation:

$$X \mapsto X \cup (X \oplus X) = X \cup \{x \oplus y | x, y \in X\}.$$

```
function [conwaygen] = conwaygen(x)
N=size(x,2);
K=N;
y=x;
for I=1:N;
    for J=1:N;
        a=conway(x(I)+x(J));
        K=K+1;
        y(K)=a;
    end;
end;
conwaygen=sort(distinct(y));
end
```

Proceeding as in Section 3.7, define the following sets of natural numbers, which can be obtained gradually using the above MATLAB codes:

$$C_0 := \{1\},$$
$$C_n := C_{n-1} \cup (C_{n-1} \oplus C_{n-1}) = C_{n-1} \cup \{x \oplus y | x, y \in C_{n-1}\}.$$

Clearly, $\mathbb{N} = \langle 1 \rangle$ is equivalent to the fact that every natural number will be covered, sooner or later, by one item (and consequently by all subsequent items) from the ascending chain $\{C_n\}_{n \geq 0}$, that is,

$$\mathbb{N} = \bigcup_{n=0}^{\infty} C_n. \tag{4.5}$$

Using **conwaygen** we obtain the first couple of terms in the sequence $\{C_n\}_{n \geq 0}$:

$\gg$ A0 = 1

1

$\gg$ A1 = conwaygen(A0)
A1 =

1   2

$\gg$ A2 = conwaygen(A1)
A2 =

1   2   3

$\gg$ A3 = conwaygen(A2)
A3 =

1   2   3   5

≫ A4 = conwaygen(A3)
A4 =

1   2   3   4   5   7

≫ A5 = conwaygen(A4)
A5 =

1   2   3   4   5   6   7   11

≫ A6 = conwaygen(A5)
A6 =

1   2   3   4   5   6   7   8   9   11   13   17

≫ A7 = conwaygen(A6)
A7 =

Columns 1 through 16

1   2   3   4   5   6   7   8   9   10   11   12   13   14   15   17

Columns 17 through 18

19   23

≫ A8 = conwaygen(A7)
A8 =

Columns 1 through 16

1   2   3   4   5   6   7   8   9   10   11   12   13   14   15   16

Columns 17 through 25

17   18   19   20   21   23   29   31   37

≫

In what follows, we will provide a proof of the remarkable cyclicity statement (4.5).

Instrumental in this derivation is a 1952 result on the distribution of primes (Nagura 1952). It is actually weaker than Nagura's result that we mentioned in Section 2.3, but for the current purposes it will suffice:

**Lemma 4.3** *For all $x \geq 8$ there exists a prime such that $x < p < 3x/2$.*

*Alternatively formulated, this ensures the existence of a prime in every interval of the form $(2y/3, y)$ for $y \geq 12$.*

*Proof of Theorem 4.2* Let $A_n$ be the maximal initial segment of natural numbers that is included in $C_n$. Let $a_n := |A_n|$ and $c_n := |C_n|$. Thus, we have

$$\{1, 2, 3, \ldots, a_n - 1, a_n\} \subseteq C_n,$$

$$a_n + 1 \notin C_n.$$

Theorem 4.2 will follow once we establish that

$$\lim_{n \to \infty} a_n = \infty. \tag{4.6}$$

Our preliminary calculation of the first couple of $C_n$'s shows that

$$(a_0, a_1, a_2, a_3, a_4, a_5, a_6) = (1, 2, 3, 3, 5, 7, 9). \tag{4.7}$$

In order to prove (4.6), we will estimate the rate of growth of the sequence $\{a_n\}_{n \geq 0}$.

From the form of the basic recurrence $C_{n+1} = C_n \cup (C_n \oplus C_n)$ and the definition of $A_n$, it follows that the set of all possible Conway sums of elements in the initial segment $A_n$,

$$A_n \oplus A_n = \{x \oplus y \mid 1 \leq x, y \leq a_n\}, \tag{4.8}$$

is a subset of $C_{n+1}$. We will estimate the size of the maximum initial segment of $\mathbb{N}$ included in the set of products (4.8).

**Lemma 4.4** *Let $p$ be any prime with*

$$p < 2a_n. \tag{4.9}$$

*Then*

$$p \in A_n \oplus A_n. \tag{4.10}$$

*Proof of Lemma 4.4* Clearly the result is vacuous for $n = 0$ [we refer to the preliminary list (4.7)]. Let $n \geq 1$. If $a_n < p < 2a_n$, then $p$ can be written as a sum $p = x + y$ with $x, y \in A_n = \{1, 2, 3, \ldots, a_n\}$. Then from (4.1) we have

$$x \oplus y = C(x + y) = C(p) = p \in A_n \oplus A_n.$$

If $p \leq a_n$, then $p \in A_n$ and $p = C(p + p) = p \oplus p \in A_n \oplus A_n$. This concludes the Proof of Lemma 4.4.

Let $q_n$ be the largest prime satisfying (4.9).

If $n \geq 5$ [which, according to the preliminary calculation (4.7), would imply $2a_n \geq 12$], setting $y := 2a_n$ in Lemma 4.3 demonstrates that that $q_n$, necessarily odd, must satisfy the estimate

$$4a_n/3 < q_n < 2a_n. \tag{4.11}$$

From Lemma 4.4, we have $q_n \in A_n \oplus A_n \subseteq C_{n+1}$. Then every product of the form

$$(2k+1) \oplus q_n, \tag{4.12}$$

where $1 \le 2k+1 \le a_n$ (that is, $2k+1$ is an odd element of $A_n$) must be an element of the set $C_{n+2}$, since

$$(2k+1) \oplus q_n \in A_n \oplus C_{n+1} \subseteq C_n \oplus C_{n+1} \subseteq C_{n+1} \oplus C_{n+1} \subseteq C_{n+2}.$$

From the definition of Conway's subprime function, since $2k+1+q_n$ is even, an element of the form (4.12) can be written as follows:

$$(2k+1) \oplus q_n = C(2k+1+q_n) = \frac{2k+1+q_n}{2} = k + \frac{1+q_n}{2}. \tag{4.13}$$

Therefore we have just found an increasing sequence of $\left\lceil \frac{a_n}{2} \right\rceil$ consecutive natural numbers, starting with $(1+q_n)/2$, that are in $C_{n+2}$.

Since $q_n < 2a_n$ (or equivalently, $(1+q_n)/2 \le a_n$), it is easy to see that the elements of the form (4.13) represent an extension of the initial segment $A_n = \{1, 2, 3, \ldots, a_n\}$ up to

$$\frac{q_n + a_n - \varepsilon_n}{2},$$

where $\varepsilon_n = 1 - (a_n \bmod 2)$. Using now the estimate (4.11) for the prime $q_n$, we find that the extension of $A_n$ by the elements of the form (4.13) actually represents an initial segment of $C_{n+2}$ whose size is bounded from below by

$$\frac{\frac{4a_n}{3} + a_n - 1}{2} = \frac{7a_n - 3}{6}.$$

Consequently, starting from $n = 5$ (corresponding to $a_5 = 7$), we have

$$a_{n+2} \ge \frac{7a_n - 1}{6}. \tag{4.14}$$

From (4.14) together with the fact that the sequence $\{a_n\}_{n \ge 0}$ is nondecreasing, (4.6) follows. This completes the Proof of Theorem 4.2.

Now comes probably the most interesting part in the exploration of the Conway subprime magma $(\mathbb{N}, \oplus)$. Unfortunately we don't have a proof for it. The idea was to look at the cardinalities of the sets in the ascending chain $\{C_n\}_{n \ge 0}$. We had previously noticed that the rate of growth appears to be exponential, but nothing prepared us for the mere possibility we were about to notice, indeed a truly unexpected sighting of the golden ratio

$$\alpha = \frac{1+\sqrt{5}}{2} = 1.61803398\ldots$$

The data acquired are displayed in the following table.

| $n$ | $|C_n|$ | $|C_n|/|C_{n-1}|$ | $n$ | $|C_n|$ | $|C_n|/|C_{n-1}|$ | $n$ | $|C_n|$ | $|C_n|/|C_{n-1}|$ |
|---|---|---|---|---|---|---|---|---|
| 1 | 2 | 2.000000000 | 12 | 138 | 1.550561798 | 23 | 25057 | 1.614081422 |
| 2 | 3 | 1.500000000 | 13 | 218 | 1.579710145 | 24 | 40442 | 1.614000080 |
| 3 | 4 | 1.333333333 | 14 | 342 | 1.568807339 | 25 | 65247 | 1.613347510 |
| 4 | 6 | 1.500000000 | 15 | 547 | 1.599415205 | 26 | 105412 | 1.615583858 |
| 5 | 8 | 1.333333333 | 16 | 882 | 1.612431444 | 27 | 170224 | 1.614844610 |
| 6 | 12 | 1.500000000 | 17 | 1429 | 1.620181406 | 28 | 274963 | 1.615301015 |
| 7 | 18 | 1.500000000 | 18 | 2299 | 1.608817355 | 29 | 444156 | 1.615330063 |
| 8 | 25 | 1.388888889 | 19 | 3705 | 1.611570248 | 30 | 717551 | 1.615538234 |
| 9 | 38 | 1.520000000 | 20 | 5961 | 1.608906883 | 31 | 1159406 | 1.615782014 |
| 10 | 56 | 1.473684211 | 21 | 9615 | 1.612984399 | 32 | 1873356 | 1.615789465 |
| 11 | 89 | 1.589285714 | 22 | 15524 | 1.614560582 | 33 | ?… | ?… |

The computation (done by Paul A. Vicol, who wrote a Julia program and let it run in a Google cloud computing environment; see Appendix B) was fairly intensive: for example, the running time for the 32nd iteration, which produced

$$C_{33} = C_{32} \cup \{x \oplus y | x, y \in C_{32}\}.$$

took 634,892 s running on the cloud, which is more than seven days. The last computed quotient satisfies

$$\frac{|C_{32}|}{|C_{31}|} - \alpha = -0.00224453\ldots$$

Let us state the conjecture thus suggested by the above calculations (Caragiu et al., to appear in Fibonacci Q.).

**The Golden Section Conjecture for Conway's Subprime Function**. Given the following:

- $C(x)$, Conway's subprime function,
- $x \oplus y = C(x+y)$, a binary operation on $\mathbb{N} = \{1, 2, 3, \ldots\}$,
- $C_0 := \{1\}$ and $C_{n+1} = C_n \cup (C_n \oplus C_n)$ for $n \geq 0$.

Then $\lim\limits_{n \to \infty} \frac{|C_{n+1}|}{|C_n|} = \frac{1+\sqrt{5}}{2}$.

If the above conjecture is true, that would provide an interesting alternative definition of the golden ratio solely in terms of prime factorization, since the recursive presentation of the sets $C_n$ is essentially built on the least prime factor function.

---

**Computational Exploration Project CEP17** Calculate values of $|C_n|$ for $n \geq 33$ and verify whether the trend toward the golden section holds.

---

## 4.5.1  On a Class of Nontrivial Finite Submagmas of $(\mathbb{N}, \oplus)$

We have seen (in Section 3.7) the important role played in the analysis of the sequences and structures based on the greatest prime factor function by the set $K_7 = \{2, 3, 5, 7\}$, which is a particularly important substructure of the groupoid, or magma, $(\Pi, *)$ of the prime numbers endowed with the binary operation *greatest prime factor of the sum*, $x * y = P(x + y)$. The set $K_7$ is the least element in the set of nontrivial structures of $(\Pi, *)$, ordered by inclusion, and also the least element in the class of substructures of $(\Pi, *)$ having the form $K_p = \{x \in \Pi | x \leq p\}$, where $p + 2 \notin \Pi$ (for example, each prime $p \equiv 1 \pmod 3$ would do).

We will now shift our attention to the Conway subprime magma $(\mathbb{N}, \oplus)$ of the positive integers with the binary operation *Conway subprime function of the sum* in order to prove the following result.

**Theorem 4.5** *For every $m \geq 2$ there exists a finite substructure of $(\mathbb{N}, \oplus)$ with $m$ elements.*

*Proof of Theorem 4.5* The proof will be based on an explicit construction of a class of finite substructures of $(\mathbb{N}, \oplus)$.

Let $A \subset \mathbb{N}$ be a finite submagma of $(\mathbb{N}, \oplus)$. From the cyclic character of $(\mathbb{N}, \oplus)$ with $\mathbb{N} = \langle 1 \rangle$ proved in the above Theorem 4.2, we must have $1 \notin A$, since otherwise, we would have $A = \mathbb{N}$.

Note that for every $a \in \mathbb{N}$ with $a \geq 2$, the 1-element set $\{a\}$ is a (trivial) submagma of $(\mathbb{N}, \oplus)$. To explore the possible structure of a nontrivial submagma $A$, assume $A = \{x_0, x_1, \ldots, x_{m-1}\}$ with $m \geq 2$ and

$$x_0 < x_1 < \cdots < x_i < x_{i+1} < \cdots < x_{m-1}. \qquad (4.15)$$

As a first observation, notice that no two nearest neighboring elements in the above list have the same parity. Indeed, if $x_j, x_{j+1}$ have the same parity, then $x_j + x_{j+1}$ is even, and because $A$ is closed under the operation $\oplus$, we have

$$x_j \oplus x_{j+1} = C(x_j + x_{j+1}) = \frac{x_j + x_{j+1}}{2} \in A.$$

However, the average value $\frac{x_j + x_{j+1}}{2}$ satisfies $x_j < \frac{x_j + x_{j+1}}{2} < x_{j+1}$, which contradicts the assumption that $x_j, x_{j+1}$ are nearest neighbors of $A$ (we shouldn't have any element of $A$ between the two).

Therefore, the integers in the list (4.15) have alternating parities. Let us now consider three elements of $A$ that appear consecutively in the list, say $x_j < x_{j+1} < x_{j+2}$. Since $x_j$ and $x_{j+2}$ have the same parity and are elements of the substructure $A$, the following must be an element of $A$:

$$x_j \oplus x_{j+2} = C(x_j + x_{j+2}) = (x_j + x_{j+2})/2.$$

But the above element, as an average, must be strictly between $x_j$ and $x_{j+2}$. Since the only such element is $x_{j+1}$, we have proved the equality

$$(x_j + x_{j+2})/2 = x_{j+1}.$$

Since this holds for every triplet of consecutively appearing elements in (4.15), we have proved that every finite submagma of $(\mathbb{N}, \oplus)$ with more than two elements is, in the arrangement (4.15), a finite arithmetic progression with an *odd* common difference $d$ (this is what makes the parities of the $x_i$'s alternate), i.e., $x_j = a + jd$ with $0 \le j \le m - 1$, so that

$$A = \{a, a + d, a + 2d, \ldots, a + (m - 1)d\}.$$

Let us first go through some small values of $m$.

### 4.5.2  Substructures with Two Elements

If $m = 2$, then $A = \{a, a + d\}$ is a submagma of $(\mathbb{N}, \oplus)$ if and only if $a \oplus (a + d) \in A$.

Let $p := LPF(2a + d)$. Then necessarily $p \ge 3$ ($d$ is odd) and

$$a \oplus (a + d) = (2a + d)/p \le (2a + d)/3 < a + d.$$

The above inequality together with $a \oplus (a + d) \in \{a, a + d\}$ implies $a \oplus (a + d) = a$, or $(2a + d)/p = a$. Thus a two-element submagma $A = \{a, a + d\}$ has $d = (p - 2)a$, in which case, for some prime $p \ge 3$, we have

$$A = \{a, (p - 1)a\}.$$

To ensure that $LPF(2a + d) = LPF(pa) = p$, we must require $LPF(a) \ge p$.

The case $m = 2$ is important because this argument made solely on $a$ and $a + d$ can be applied to each of the cases with $m \ge 2$. Thus in each such case, we know offhand that the common difference in the arithmetic progression $A$ is

$$d = (p - 2)a,$$

with $p \geq 3$ prime and $LPF(a) \geq p$.

Consequently, since the common difference $d$ is odd, it follows that the least element $a \in A$ is odd, and the expected presentation of the elements of the sub-magma $A$ is

$$A = \{a[1 + i(p - 2)] | 0 \leq i \leq m - 1\} = \{(ip - (2i - 1))a | 0 \leq i \leq m - 1\}.$$

More explicitly hands-on, the elements of $A$ should be, in increasing order, beginning with the odd integer $a$,

$$a < (p - 1)a < (2p - 3)a < (3p - 5)a < (4p - 7)a < \cdots < ((m - 1)p - (2m - 3))a$$

### 4.5.3   Substructures with Three Elements

If $m = 3$ we expect $A$ to have the following structure:

$$A = \{a, (p - 1)a, (2p - 3)a\}.$$

In the analysis of the first two elements, we already required $LPF(a) \geq p$. To have closure under $\oplus$, it will be enough to require

$$(p - 1)a \oplus (2p - 3)a = C((3p - 4)a) = \frac{(3p - 4)a}{LPF((3p - 4)a)} \in A.$$

Since $LPF((3p - 4)a) \geq 3$, we have $\frac{(3p-4)a}{LPF((3p-4)a)} < (p - 1)a$ and thus we have $\frac{(3p-4)a}{LPF((3p-4)a)} = a$, so that $LPF((3p - 4)a) = 3p - 4$. This implies that $3p - 4$ is a prime and $LPF(a) \geq 3p - 4$. Since for an odd prime $p$ we have $3p - 4 > p$, it follows that the new requirement $LPF(a) \geq 3p - 4$ supersedes the previous one $(LPF(a) \geq p)$, and we can conclude that the substructures with three elements of $(\mathbb{N}, \oplus)$ are of the form $A = \{a, (p - 1)a, (2p - 3)a\}$ with $3p - 4$ a prime and $LPF(a) \geq 3p - 4$.

Let's see some examples.

- If $p = 3$, we are looking at $A = \{a, 2a, 3a\}$ with $LPF(a) \geq 5$. For instance, $A$ could be $\{5, 10, 15\}$, or $\{7, 14, 21\}$, etc.
- If $p = 5$, we are looking at $A = \{a, 4a, 7a\}$ with $LPF(a) \geq 11$. For instance, $A$ could be $\{11, 44, 77\}$, or $\{13, 52, 91\}$, etc.

### 4.5.4   Concluding the Proof of Theorem 4.5

There is, however, a fortunate setup that will allow us to produce, for each $m \geq 2$, a substructure of $(\mathbb{N}, \oplus)$ with $m$ elements, thus concluding the Proof of Theorem 4.5.

In light of the arithmetic progression structure, let us look for $A$ of the following simple form:

$$A = \{a, 2a, 3a, 4a, \ldots, ma\}.$$

Here $a \geq 3$ is an odd integer, subject to constraints that are to be determined.

Clearly the basic requirement is to have $A$ closed under the Conway sum operation $\oplus$. That is, for every $k, l$ with $1 \leq k \leq l \leq m$ we need to have $k$, $l$ $(ka) \oplus (la) \in A$. But

$$(ka) \oplus (la) = C((k+l)a).$$

The best choice for $a$ will be the inspired by the requirement of having $C((k+l)a) \in A$, after noticing that $k + l \leq 2m - 1$ for $k \neq l$ (if $k = l$ there is nothing to prove, since $(ka) \oplus (la) = ka \in A$).

To that end, it will be enough to choose $a$ such that all prime factors of $A$ are greater than or equal to $2m - 1$, that is,

$$LPF(a) \geq 2m - 1.$$

Once the above is satisfied, noticing that $(k + l)a$ is obviously composite, we see that the result of the operation $(ka) \oplus (la)$ for $k \neq l$ is an element of $A = \{a, 2a, 3a, 4a, \ldots, ma\}$, since then, due to the choice of $a$, $LPF((k+l)a) = LPF(k+l)$,

$$(ka) \oplus (la) = \frac{(k+l)a}{LPF((k+l)a)} = \left[\frac{(k+l)}{LPF((k+l)a)}\right]a,$$

and $1 \leq \frac{(k+l)}{LPF(k+l)} \leq \frac{k+l}{2} \leq m$. The Proof of Theorem 4.5 is now complete. Just to provide a numerical example, let's say $m = 7$. As a substructure of $(\mathbb{N}, \oplus)$ with seven elements we can take $a = 13$ (such that its least prime factor is at least $2m - 1$), and thus

$$A = \{13, 26, 39, 52, 65, 78, 91\}.$$

Let us build the Conway addition table for the above 7-element substructure.

| $\oplus$ | 13 | 26 | 39 | 52 | 65 | 78 | 91 |
|----------|----|----|----|----|----|----|----|
| 13 | 13 | 13 | 26 | 13 | 39 | 13 | 52 |
| 26 | 13 | 26 | 13 | 39 | 13 | 52 | 39 |

(continued)

(continued)

| $\oplus$ | 13 | 26 | 39 | 52 | 65 | 78 | 91 |
|---|---|---|---|---|---|---|---|
| 39 | 26 | 13 | 39 | 13 | 52 | 39 | 65 |
| 52 | 13 | 39 | 13 | 52 | 39 | 65 | 13 |
| 65 | 39 | 13 | 52 | 39 | 65 | 13 | 78 |
| 78 | 13 | 52 | 39 | 65 | 13 | 78 | 13 |
| 91 | 52 | 39 | 65 | 13 | 78 | 13 | 91 |

### 4.5.5 On a Class of Nontrivial Infinite Submagmas of $(\mathbb{N}, \oplus)$

This is a problem that recently emerged as a result of a discussion with a group of students: clearly $\mathbb{N} = \{1, 2, 3, \ldots\}$ is a submagma of $(\mathbb{N}, \oplus)$. Let's see what can be said, then (in the matter of being a submagma or not) about sets of natural numbers of the form

$$\mathbb{N}^{\geq n} := \{n, n+1, n+2, \ldots\}.$$

This turned out to be an interesting, though easy, elementary problem. Let us denote by $M$ the set of natural numbers $n$ such that $\mathbb{N}^{\geq n}$ is a submagma of $(\mathbb{N}, \oplus)$. Clearly $1 \in M$.

We will begin by considering a few small values of $n$. While we discover at the beginning some elements of $M$, we will see that the elements of $M$ appear to become extinct as we advance.

- $2 \in M$: Indeed, if $x, y \in \{2, 3, 4, \ldots\}$ then $x \oplus y = C(x+y)$ can never be 1 (in fact, the definition of $C$ ensures that $C(x) = 1$ if and only if $x = 1$).
- $3 \in M$: since if $x, y \in \{3, 4, 5, \ldots\}$, then $x + y \geq 6$, in which case it is easy to see that $C(x+y) \geq 3$.
- $4 \notin M$: since $4, 5 \in \mathbb{N}^{\geq 4}$ but $4 \oplus 5 = (4+5)/3 = 3 \notin \mathbb{N}^{\geq 4}$.
- $5 \in M$: since it is easy to see that if $x, y \in \{5, 6, 7, \ldots\}$ then $C(x+y)$ cannot be 1, 2, 3, or 4.
- $6 \notin M$: since $6, 9 \in \mathbb{N}^{\geq 6}$ but $6 \oplus 9 = (6+9)/3 = 5 \notin \mathbb{N}^{\geq 6}$.
- $7 \notin M$: same as before, counterexample $7 \oplus 8 = 5 \notin \mathbb{N}^{\geq 7}$.
- $8 \notin M$: counterexample $8 \oplus 13 = 7 \notin \mathbb{N}^{\geq 8}$.
- $9 \notin M$: counterexample $9 \oplus 12 = 7 \notin \mathbb{N}^{\geq 9}$.
- $10 \notin M$: counterexample $10 \oplus 11 = 7 \notin \mathbb{N}^{\geq 10}$.
- $11 \notin M$: counterexample $11 \oplus 14 = 5 \notin \mathbb{N}^{\geq 11}$.
- $12 \notin M$: counterexample $12 \oplus 13 = 5 \notin \mathbb{N}^{\geq 12}$.
- $13 \notin M$: counterexample $13 \oplus 14 = 9 \notin \mathbb{N}^{\geq 13}$.

- $14 \notin M$: counterexample $16 \oplus 17 = 11 \notin \mathbb{N}^{\geq 14}$. Note that the same counterexample can be used to show that $15 \notin M$ and $16 \notin M$.
- $17 \notin M$: counterexample $17 \oplus 18 = 7 \notin \mathbb{N}^{\geq 17}$.

Having done the above verifications, let us attempt to prove that there will be no further elements of $M$ that will subsequently emerge.

The idea is simple: if $n$ is large, then $\mathbb{N}^{\geq n} = \{n, n+1, n+2, \ldots\}$ is expected not to be a submagma, because we will find two small elements of $\mathbb{N}^{\geq n}$ (small in the sense of being close to $n$) with odd sum divisible by 3, in which case dividing their sum by the least prime factor of the sum (which would be 3) will bring us below $n$. This idea suggests a proof by cases given by the congruence class of $n \geq 18$ modulo 6. The six cases will be $n = 6k + r (k \geq 3)$ with remainder $r = 0, 1, \ldots, 5$:

- $n = 6k$. Then $\mathbb{N}^{\geq n} = \{6k, 6k+1, 6k+2, 6k+3, 6k+4, 6k+5, \ldots\}$ is not a submagma of $(\mathbb{N}, \oplus)$ since $(6k) \oplus (6k+3) = (12k+3)/3 = 4k+1 < 6k$.
- $n = 6k+1$. Then $\mathbb{N}^{\geq n} = \{6k+1, 6k+2, 6k+3, 6k+4, 6k+5, 6k+6, \ldots\}$ is not a submagma of $(\mathbb{N}, \oplus)$ since $(6k+1) \oplus (6k+2) = (12k+3)/3 = 4k+1 < 6k+1$.
- $n = 6k+2$. Then $\mathbb{N}^{\geq n} = \{6k+2, 6k+3, 6k+4, 6k+5, 6k+6, 6k+7, \ldots\}$ is not a submagma of $(\mathbb{N}, \oplus)$ since $(6k+4) \oplus (6k+5) = (12k+9)/3 = 4k+3 < 6k+2$.
- $n = 6k+3$. Then $\mathbb{N}^{\geq n} = \{6k+3, 6k+4, 6k+5, 6k+6, 6k+7, 6k+8, \ldots\}$ is not a submagma of $(\mathbb{N}, \oplus)$ since $(6k+4) \oplus (6k+5) = (12k+9)/3 = 4k+3 < 6k+3$.
- $n = 6k+4$. Then $\mathbb{N}^{\geq n} = \{6k+4, 6k+5, 6k+6, 6k+7, 6k+8, 6k+9, \ldots\}$ is not a submagma of $(\mathbb{N}, \oplus)$ since $(6k+4) \oplus (6k+5) = (12k+9)/3 = 4k+3 < 6k+4$.
- $n = 6k+5$. Then $\mathbb{N}^{\geq n} = \{6k+5, 6k+6, 6k+7, 6k+8, 6k+9, 6k+10, \ldots\}$ is not a submagma of $(\mathbb{N}, \oplus)$ since $(6k+7) \oplus (6k+8) = (12k+15)/3 = 4k+5 < 6k+5$.

Thus we have proved that the only values of $n$ for which $\mathbb{N}^{\geq n} = \{n, n+1, n+2, \ldots\}$ is a submagma of $(\mathbb{N}, \oplus)$ are 1, 2, 3, and 5.

We end this chapter with a few words on commuting pairs, returning to the "prologue" of Section 3.7, where we have seen that in the GPF magma $(\Pi, f_{2,1})$ on the set of primes $\Pi$ with $f_{2,1}(x, y) = P(2x + y)$ the only commuting pairs $(x, y)$ with $x < y$ are $(2, 5)$ and $(2, 23)$.

Now, if we switch to using Conway's subprime function instead of the greatest prime factor function, we may wish to search, in this new context, for pairs $(x, y)$ of natural numbers satisfying the commutation relation

$$C(x + 2y) = C(2x + y).$$

For example, a little trial and error shows that if $x$ is odd and $y = 4x$, then $(x, y)$ is a commuting pair, since

$$C(x+2y) = C(x+8y) = C(9x) = 9x/3 = 3x = 6x/2 = C(6x) = C(2x+4x)$$
$$= C(2x+y).$$

For more, we invite the inquisitive student to join us in the excitement of experimentation with the following "commuting pairs project."

**Computational Exploration Project CEP18** For two distinct positive integers $a$, $b$, look for pairs (and patterns of pairs) of natural numbers $(x, y)$ that "commute" in the sense that $C(ax+by) = C(ay+bx)$, where $C$ is Conway's subprime function.

# Chapter 5
# Going All Experimental: More Games and Applications

In the first part of this chapter we will explore new types of cellular automata that emerged from our work with Ducci games based on the greatest prime factor or Conway's subprime function.

The quaternary cellular automata are essentially based on the fact that the set of the first four primes, 2, 3, 5, and 7, forms a closed (nonassociative) structure under the "greatest prime factor of the sum" algebraic operation. We will see that every nonassociative word can be used to define such an automaton (in fact, the construction can be adapted to any number of dimensions).

We will move on to a class of automata essentially based on multidimensional greatest prime factor sequences for which the local states of the cells in the grid are integers (finitely many if the ultimate periodicity conjecture for such sequences is valid). We will be able to witness states of quite amazing complexity in their evolution.

In the last two sections of the chapter we will explore various practical ways of generating random sequences and pseudorandom (or so we thought) walks (in 1D, 2D, and beyond) using greatest prime factor sequences. We discover two amazing things:

- In Section 5.3 we discover that the $\pm 1$ walks associated with greatest prime factor sequences have a curious large-scale tendency toward a negative displacement. This shows that while we cannot use them globally as genuine random walk models (though some limited-length segments can be used as such), the "real deal" is a possibly deep connection with a phenomenon resembling the celebrated Chebyshev bias and a related conjecture in the area.
- In Section 5.4 we define a class of bitstreams (0–1 sequences) based on greatest prime factor sequences and discover that they generally behave very well from the points of view of quasirandomness (in a representation as subsets as residue class rings) and linear complexity, thus making the case for a nice potential application to cryptography (stream cipher).

© Springer International Publishing AG 2017
M. Caragiu, *Sequential Experiments with Primes*,
DOI 10.1007/978-3-319-56762-4_5

The topics explored in both Sections 5.3 and 5.4 offered pleasant surprises for us (one, related to Chebyshev-like bias, being sort of "theoretical"; the other, related to the discovery of sequences with good linear complexities, being "practical").

## 5.1    The Greatest Prime Factor and "Nonassociative" Quaternary Cellular Automata

Beginning in the 1940s with the pioneering ideas of Stanislaw Ulam (involving processes of crystal growth) and John von Neumann (involving self-replicating robots), the exploration of cellular automata has gained extraordinary prominence and is today one of the classical themes in experimental and computational mathematics.

Among the reasons cellular automata are studied today, one may list, following (Ilachinski 1993), their power as computational engines, as discrete dynamical systems simulators, as theoretical facilitators in understanding pattern formation and complexity, and as original models of theoretical physics.

In the classical cellular automaton setup, the nature of the local states is binary (i.e., each cell hosts a local variable in $\mathbb{Z}/2\mathbb{Z}$).

Of particular interest (especially due to our work on Ducci iteration in various settings) is "rule 102," in which each state at a subsequent moment is the sum of the local states of the same cell and its right neighbor at the previous moment. Therein lies the reason for the number assigned to the rule. The following table shows the state of each cell at moment $t + 1$ in terms of the states of the three cells of its Moore neighborhood of range 1 at moment $t$:

| 111 | 110 | 101 | 100 | 011 | 010 | 001 | 000 |
|-----|-----|-----|-----|-----|-----|-----|-----|
| 0   | 1   | 1   | 0   | 0   | 1   | 1   | 0   |

The first row represents the eight possible states in such a Moore neighborhood (in decreasing order), while 102 represents the binary encoding of the bits in the second row.

The following grid is a fragment in the evolution of the CA rule 102 starting from a state (let's call it "row 0") that is infinite and has one cell hosting a 1.

| ... | 0 | 0 | 0 | 0 | 0 | 0 | 0 | 0 | 1 | 0 | ... |
|-----|---|---|---|---|---|---|---|---|---|---|-----|
| ... | 0 | 0 | 0 | 0 | 0 | 0 | 0 | 1 | 1 | 0 | ... |
| ... | 0 | 0 | 0 | 0 | 0 | 0 | 1 | 0 | 1 | 0 | ... |
| ... | 0 | 0 | 0 | 0 | 0 | 1 | 1 | 1 | 1 | 0 | ... |
| ... | 0 | 0 | 0 | 0 | 1 | 0 | 0 | 0 | 1 | 0 | ... |
| ... | 0 | 0 | 0 | 1 | 1 | 0 | 0 | 1 | 1 | 0 | ... |
| ... | 0 | 0 | 1 | 0 | 1 | 0 | 1 | 0 | 1 | 0 | ... |

(continued)

(continued)

| ... | 0 | 1 | 1 | 1 | 1 | 1 | 1 | 1 | 1 | 0 | ... |
|---|---|---|---|---|---|---|---|---|---|---|---|
| ... | 1 | 0 | 0 | 0 | 0 | 0 | 0 | 0 | 1 | 0 | ... |
| ... | ... | ... | ... | ... | ... | ... | ... | ... | ... | ... | ... |

Note that if we replace addition modulo 2 with the regular addition of integers, we will be facing a left-tilted Pascal triangle in which the nonzero elements in the $n$th row are $\binom{n}{0}, \binom{n}{1}, \binom{n}{2}, \binom{n}{3}, \ldots, \binom{n}{n}$. Incidentally, comparing the two tables reveals the classical result to the effect that all binomial coefficients $\binom{n}{k}$ are odd if and only if $n = 2^k - 1$ for some $k \geq 1$.

| ... | 0 | 0 | 0 | 0 | 0 | 0 | 0 | 0 | 1 | 0 | ... |
|---|---|---|---|---|---|---|---|---|---|---|---|
| ... | 0 | 0 | 0 | 0 | 0 | 0 | 0 | 1 | 1 | 0 | ... |
| ... | 0 | 0 | 0 | 0 | 0 | 0 | 1 | 2 | 1 | 0 | ... |
| ... | 0 | 0 | 0 | 0 | 0 | 1 | 3 | 3 | 1 | 0 | ... |
| ... | 0 | 0 | 0 | 0 | 1 | 4 | 6 | 4 | 1 | 0 | ... |
| ... | 0 | 0 | 0 | 1 | 5 | 10 | 10 | 5 | 1 | 0 | ... |
| ... | 0 | 0 | 1 | 6 | 15 | 20 | 15 | 6 | 1 | 0 | ... |
| ... | 0 | 1 | 7 | 21 | 35 | 35 | 21 | 7 | 1 | 0 | ... |
| ... | 1 | 8 | 28 | 56 | 70 | 56 | 28 | 8 | 1 | 0 | ... |
| ... | ... | ... | ... | ... | ... | ... | ... | ... | ... | ... | ... |

The GPF Ducci games (for which the limit behavior was discussed in Section 3.6) can be viewed as nonassociative analogues of the 1-dimensional elementary additive cellular automaton, known under the label "rule 102" under Wolfram's encoding system (Weisstein; Wolfram 1983; Martin et al. 1984).

Once they enter the limit cycle, the behavior of the classical Ducci game is essentially binary. On the other hand, once they enter the limit cycle, the behavior of the GPF Ducci games is essentially quaternary. The following table recalls the GPF Ducci iteration, where the nonassociative operation "*" is defined by $x * y = P(x+y)$:

| ... | $x_{n-1}$ | $x_n$ | $x_{n+1}$ | ... |
|---|---|---|---|---|
| ... | $x_{n-1} * x_n$ | $x_n * x_{n+1}$ | $x_{n+1} * x_{n+2}$ | ... |

A complete account of the limit periods for GPF Ducci games in dimensions up to 8 can be found in Section 3.6.

This evolution rule for the GPF Ducci games can be generalized. To this end, we will set up a "nonassociative classification system" for quaternary cellular automata for which the possible states per cell are 2, 3, 5, and 7 and the iteration uses the operation "*".

The "code" for such a generalized quaternary CA in dimension $d$ will be a nonassociative word formed with variables $x_k (k \in \mathbb{Z})$ and "$*$".

Let's say we would like to define the state of a cell at the moment $t + 1$. Due to cyclicity we may assume that the cell of interest is labeled "cell 0". Expressing the state of "cell 0" at the moment $t + 1$ in terms of the states a set of cells at moment $t$ is the same as specifying a non-associative word in the variables $x_k (k \in \mathbb{Z})$.

For example, the word corresponding to the GPF Ducci game (which we view as an analogue of Wolfram's "rule 102") is

$$w = x_0 * x_1.$$

That is, the state of the "center" (or "cell 0") at moment $t + 1$ is the $*$ product between the state of cell 0 at moment $t$ and the state of cell 1 at moment $t$ as encoded by the above word $w$. But due to cyclicity, the center can be anywhere so the code $w = x_0 * x_1$ symbolizes that at each moment, $x_n(t+1) = x_n(t) * x_{n+1}(t)$, with the subscripts being taken (cyclically) modulo the dimension $d$.

To see how these codes work in practice, let us take, for example, the case $d = 5$ and

$$w = (x_{-1} * x_0) * x_2.$$

That is, for every $0 \le n \le 4$ the state $x_n$ of cell $n$ at moment $t + 1$ is expressed as the product $(x_{n-1} * x_n) * x_{n+2}$ involving the states, at moment $t$, of the cells $n - 1, n, n + 2$, with the subscripts seen modulo 5, which makes, for example, $x_0 = x_5, x_{-1} = x_4$, etc.

The following table makes this rule explicit, where the first row represents moment $t$, while the second row represents moment $t + 1$:

| $x_1$ | $x_2$ | $x_3$ | $x_4$ | $x_5$ |
| --- | --- | --- | --- | --- |
| $(x_5 * x_1) * x_3$ | $(x_1 * x_2) * x_4$ | $(x_2 * x_3) * x_5$ | $(x_3 * x_4) * x_1$ | $(x_4 * x_5) * x_1$ |

To implement this in MAPLE, we first use, for technical reasons, a procedure (which will be applied to subscripts) that performs the mod 5 shift from the scale 0, 1, 2, 3, 4 to the scale 1, 2, 3, 4, 5 respectively:

```
modup:=proc(a::integer, n::integer)::integer;
if a mod n <>(n-1) then modup:=1+(a mod n) else modup:=n; end if
end proc;
```

The above-mentioned quaternary cellular automaton rule was implemented as follows:

```
CADucci:= proc(x::list)::list;
local d, k, X;
d:=numelems(x);
for k from 1 to d do X[k]:=P(P(x[modup(k-1,d)]
+x[modup(k,d)])+x[modup(k+2,d)]) end do;
[seq(X[k],k=1..d)];
end proc;
```

For example, if we start with [2, 3, 5, 5, 7], the evolution encoded by the nonassociative word $(x_{-1} * x_0) * x_2$ will produce [5, 3, 7, 3, 2], then [5, 7, 5, 2, 7], then [5, 5, 3, 5, 2], etc.

Is there any real difference between the statistics of limit periods of the quaternary automaton encoded by $(x_{-1} * x_0) * x_2$ and that encoded by $x_0 * x_1$ (which is precisely the "cycle mode" of Section 3.6 GPF Ducci game)?

An exploratory Floyd–Monte Carlo analysis of the periods of the quaternary automaton encoded by $(x_{-1} * x_0) * x_2$ shows that a difference indeed exists. Note that

```
> ithprime(rand(1..4)());
```

produces a random element in $K_7 = \{2,3,5,7\}$.

```
> M:=1000: for r from 1 to M do
A1:=ithprime(rand(1..4)()); A2:=ithprime(rand(1..4)());
A3:=ithprime(rand(1..4)()); A4:=ithprime(rand(1..4)());
A5:=ithprime(rand(1..4)()); X:=[A1,A2,A3,A4,A5];
SLOW:=CADucci(X): FAST:=CADucci(CADucci(X)): k:=1: while
SLOW<>FAST do SLOW:=CADucci(SLOW):
FAST:=CADucci(CADucci(FAST)): k:=k+1: end do:
TENTATIVE_START:=k; X:=SLOW: SLOW:=CADucci(X):
FAST:=CADucci(CADucci(X)): k:=1: while SLOW<>FAST do
SLOW:=CADucci(SLOW): FAST:=CADucci(CADucci(FAST)): k:=k+1:
end do: PERIOD[r]:=k; end do:
PER:=[seq(PERIOD[r],r=1..M)]:
CYCLE_LENGTHS:=convert(PER,set);
OCCURRENCES:=seq(Occurrences(CYCLE_LENGTHS[t],PER),t=1..nu
melems(CYCLE_LENGTHS));
```

As a result, we find limit cycles of lengths 1, 15, and 30. A typical output is more or less something like this:

```
CYCLE_LENGTHS:= {1, 15, 30}
OCCURRENCES:=2, 401, 597
```

On the other hand, for the CA encoded by $x_0 * x_1$, as indicated in Section 3.6, the lengths of all possible limit cycles are 1, 20, 30, 40.

### 5.1.1 Two-Dimensional Nonassociative Quaternary Cellular Automata

Quaternary CA rules of evolution specified by a nonassociative keyword with the "*" operation can be extended to any dimension. For example, in 2D we need only a nonassociative word

$$w = w(x_{k_1,l_1}, x_{k_2,l_2}, \ldots, x_{k_r,l_r})$$

in *doubly-indexed* variables $x_{k,l}$ with $k, l \in \mathbb{Z}$. Then the state

$$x_{m,n}(t+1)$$

of the site $(m, n)$ at moment $t + 1$ can be expressed in terms of a set of states at moment $t$ as follows:

$$x_{m,n}(t+1) = w\big(x_{m+k_1,n+l_1}(t), x_{m+k_2,n+l_2}(t), \ldots, x_{m+k_r,n+l_r}(t)\big).$$

Let's take a concrete example, say

$$w = \big(x_{-1,0} * x_{1,0}\big) * \big(x_{0,-1} * x_{0,1}\big).$$

In easy to visualize terms, the rules of the game are as follows:

- The evolution occurs on an $N \times N$ grid, with periodic boundary conditions.
- There are only four possible local variables, the elements of the set $K_7 := \{2, 3, 5, 7\}$.
- The local variable at every cell $C$ is updated with the following algebraic formula expressed solely in terms of *, which involves the local variables hosted by its four cardinal point neighbors, as $C \mapsto (N * S) * (E * W)$. Recall that $K_7 := \{2, 3, 5, 7\}$ is closed under the operation *.

$$
\begin{array}{c}
N \\
E \quad \boxed{C} \quad W \\
S
\end{array}
$$

Note that if we replace $K_7$ with another submagma $X$ of $(\Pi, *)$, we will obtain a similar automaton with the set of possible local states $X$.

But the $K_7$ setting defines a quaternary cellular automaton, and it can be projected onto a binary one by choosing a rule such as mapping 2 and 5 to 1, while 3 and 7 are mapped to 0.

In the process, Paul A. Vicol, seduced by these "prime games," built applets that offer a dynamical visualization of the evolution of this binary automaton (in black and white) as well as well as that of its quaternary parent (in color). Screen snapshots from these two are shown below.

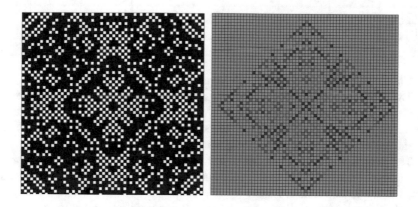

## 5.2   Complex Evolution for a Class of Integer-Valued Nonassociative Automata

For another variation, let's build a GPF analogue of the 1-dimensional cellular automaton that is recorded as "rule 150" under Wolfram's classification, that is, its local evolution is condensed in the following table:

| 111 | 110 | 101 | 100 | 011 | 010 | 001 | 000 |
|-----|-----|-----|-----|-----|-----|-----|-----|
| 1   | 0   | 0   | 1   | 0   | 1   | 1   | 0   |

For the greatest prime factor analogue, the local variable assigned to cell $k$ at moment $t + 1$ is the greatest prime factor of the sum of the local variables of the cells $k - 1, k$, and $k + 1$ at moment $t$:

$$x_n(t + 1) = P(x_{n-1}(t) + x_n(t) + x_{n+1}(t)).$$

Let us assume that we start from a "seed" consisting of a row with nine cells, with 2 in each cell but one, which hosts a 3. Repeated application of the recurrence rule will produce the following development:

| 2 | 2  | 2  | 2  | 3  | 2  | 2  | 2  | 2 |
|---|----|----|----|----|----|----|----|---|
| 3 | 3  | 3  | 7  | 7  | 7  | 3  | 3  | 3 |
| 3 | 3  | 13 | 17 | 7  | 17 | 13 | 3  | 3 |
| 3 | 19 | 11 | 37 | 41 | 37 | 11 | 19 | 3 |

(continued)

(continued)

| 5 | 11 | 67 | 89 | 23 | 89 | 67 | 11 | 5 |
|---|---|---|---|---|---|---|---|---|
| 7 | 83 | 167 | 179 | 67 | 179 | 167 | 83 | 7 |
| 97 | 257 | 13 | 59 | 17 | 59 | 13 | 257 | 97 |
| 41 | 367 | 47 | 89 | 5 | 89 | 47 | 367 | 41 |
| 449 | 13 | 503 | 47 | 61 | 47 | 503 | 13 | 449 |

Clearly this game is no longer quaternary!

The states of this (nontraditional, nonassociative) cellular automaton are integers, specifically primes. They can get very large, but under reasonable ultimate periodicity assumptions for vector-valued greatest prime factor sequences, one might venture to say that only a finite number of states can appear.

The evolution of this game looks like a curious "roller coaster": large primes appear to emerge in the process, but they are sometimes suddenly followed by large drops in the local variable values.

We will consider similar types of such 2-dimensional analogues of classical 2-dimensional cellular automata. What we are actually looking for is the evolution of states on a finite grid (we may consider models with or without periodic boundary conditions) in which every cell hosts a local variable with prime values.

This exploration was done, as a joint project with Paul A. Vicol (then a student at Simon Fraser) in the fall of 2014. In our model, every cell updates as the greatest prime factor of an integer combination with nonnegative coefficients (not all zero) of the values assigned to the cells in its Moore neighborhood (or an extended Moore neighborhood); see, e.g., (Weisstein):

$$x_{m,n} \mapsto P\big(Zx_{m,n} + Ax_{m-1,n+1} + Bx_{m,n+1} + Cx_{m+1,n+1} + Dx_{m+1,n}$$
$$+ Ex_{m+1,n-1} + Fx_{m,n-1} + Gx_{m-1,n-1} + Hx_{m-1,n}\big).$$

If we assume periodic boundary conditions, that is, that every cell is an "inner cell," then all cells satisfy the same recurrence. If we do not assume periodic boundary conditions, then the recurrence rule for the boundary cells should be defined independently of the recurrence obeyed by the "inner" cells.

Just to see the evolution in a concrete example, let's say we use the following model:

$$Z = 0$$
$$A = B = C = D = E = F = G = H = 1$$

In this setup, the value at every cell updates as the greatest prime factor of the values assigned to all the cells in its Moore neighborhood except the center:

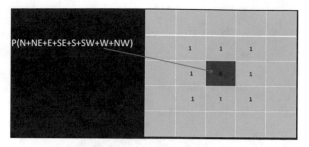

With this iteration rule, on a 7-by-7 grid, if we start from the initial state

$$
\begin{bmatrix}
2 & 2 & 2 & 2 & 2 & 2 & 2 \\
2 & 2 & 2 & 2 & 2 & 2 & 2 \\
2 & 2 & 2 & 2 & 2 & 2 & 2 \\
2 & 2 & 2 & 2 & 2 & 2 & 2 \\
2 & 2 & 2 & 2 & 2 & 2 & 2 \\
2 & 2 & 2 & 2 & 2 & 2 & 2 \\
2 & 2 & 2 & 2 & 2 & 2 & 2
\end{bmatrix},
$$

we will arrive, after eight steps, at the following configuration:

$$
\begin{bmatrix}
11 & 409 & 19 & 229 & 19 & 409 & 11 \\
409 & 43 & 61 & 829 & 61 & 43 & 409 \\
19 & 61 & 257 & 89 & 257 & 61 & 19 \\
229 & 829 & 89 & 7 & 89 & 829 & 229 \\
19 & 61 & 257 & 89 & 257 & 61 & 19 \\
409 & 43 & 61 & 829 & 61 & 43 & 409 \\
11 & 409 & 19 & 229 & 19 & 409 & 11
\end{bmatrix}.
$$

The following is a MATLAB "**bar3**-style" display of the configuration at "$T = 81$" on a 41-by-41 grid starting from a similar seed configuration (with a 3 in the center and 2's everywhere else) and subject to the same recurrence relation, with every cell updated as the greatest prime factor of the values of the eight cells surrounding it).

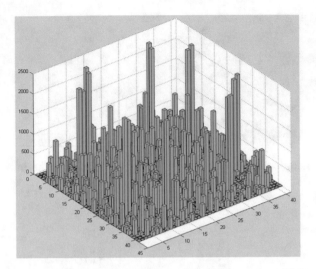

Note that the recurrence forms involved in all these types of 2D "nonassociative GPF cellular automata" where local states are synchronously updated as the greatest prime factor of a weighted sum of the variables corresponding to the cells in a certain extended Moore neighborhood (possibly extending to the whole grid in a cyclic fashion if we assume periodic boundary conditions) are special cases of the vector-valued multidimensional greatest prime factor sequences (MGPF sequences) discussed in Section 3.3, and hence, under the MGPF conjecture, are believed ultimately to enter a limit cycle.

## 5.2.1   Taking the Boundary into Account

Let us give up the hypothesis of periodic boundary conditions, so that the grid abruptly ends at the boundary. Let us assume, for simplicity, that for interior cells that do not have common points with the boundary, the recurrence relation has the same form as in the previous example, so that an interior cell updates as the greatest prime factor of the sum of the eight cells around itself.

The recurrence forms for the boundary cells need to be defined separately. Let us assume, for simplicity, that the boundary cells update as the greatest prime factor of the sum of the three Moore neighbors (for the corners) and five Moore neighbors (for boundary cells that are not corners), respectively:

Let us choose this recurrence setup on a 27-by-27 grid with the initial configuration having a 3 in the center and 2's everywhere else, and look for the boundary effects. For more variation we will change the matrix visualization style, using MATLAB's **pcolor**, and see what happens at various moments (0, 3, 5, 7, 10, and 17 are shown below). We notice an interesting wavelike development at the boundary in the initial time interval, independent of the development at the center, with the two developments interacting later on.

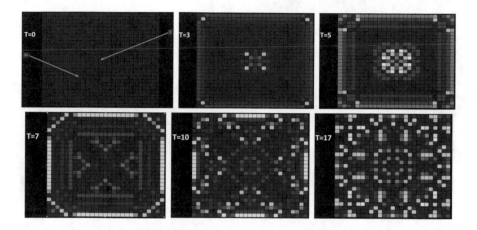

Let's now keep the form of the recurrence at the boundary and modify the recurrence formula for the inner cells in a way that will be slightly asymmetric (or one could say "almost symmetric" with a "southwest tilt"):

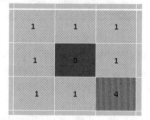

The southwest tilt reflects in the shape of the initial developments, both at the center and along the boundary:

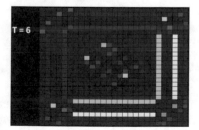

Finally, having the boundary cells undergo the same recurrence as before, let's add an additional dose of asymmetry to the recurrence for the inner cells:

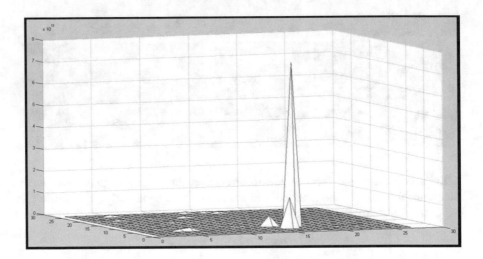

For the sake of variety of perspective, let us use MATLAB's **mesh** display to visualize the state of this automaton just after the 31st iteration.

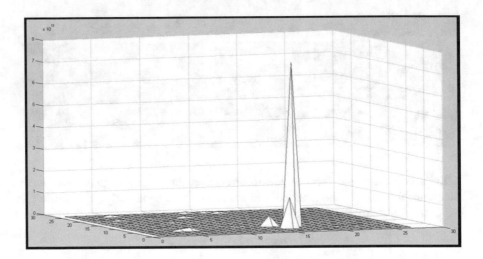

Note that the local variable hosted at one particular moment is particularly large compared to all the others. Again, the truth of the MGPF conjecture implies the existence of an upper bound, however high, for the values assigned to the cells in the whole iterative process.

In our experiments with various settings in the iteration rules, these nonassociative GPF cellular automata were found typically to evolve in an apparently unpredictable "roller coaster" way. They seem interesting and attractive from a complex systems point of view.

### 5.2.2    MATLAB for 2D GPF Automata

In exploring 2-dimensional systems, we used a few MATLAB functions. The functions listed below can be modified if one wishes to change the setup and assign different weights to the nine cells in the Moore surrounding. The update of the inner cells is based on the setup

$$Z = 0$$
$$A = B = D = F = G = H = 1$$
$$C = 7$$
$$E = 4$$

while the boundary cells are updated based on three weights of 1 for the cells surrounding the corners and five weights of 1 for the cells surrounding noncorner cells.

The above are the coefficients we used back in the fall of 2014. Experimental data obtained subsequently showed that the choice $Z = 0$ doesn't appear to make any particular difference when compared to the case $Z \neq 0$, that is, it does not have any particular impact on the overall qualitative evolutionary aspects of the system. Further on, we will present an example of an automaton with $Z \neq 0$.

Recall that everything will be based on the greatest prime factor:

```
function gpf=gpf(n)
gpf=max(factor(n));
```

The next function represents just one step in the evolution of the automaton:

```
function [ gpf2dim ] = gpf2dim(A)
N=size(A,2);
for I=2:N-1;
    for J=2:N-1;
        B(I,J)=gpf(A(I,J+1)+4*A(I-1,J+1)+7*A(I+1,J+1)+A(I-
1,J)+A(I+1,J)+A(I-1,J-1)+A(I,J-1)+A(I+1,J-1));
    end;
end;
for J=2:N-1;
    B(1,J)=gpf(A(1,J-1)+A(1,J+1)+A(2,J-1)+A(2,J)+A(2,J+1));
end;
for J=2:N-1;
    B(N,J)=gpf(A(N,J-1)+A(N,J+1)+A(N-1,J-1)+A(N-1,J)+A(N-1,J+1));
end;
for I=2:N-1;
    B(I,1)=gpf(A(I-1,1)+A(I+1,1)+A(I-1,2)+A(I,2)+A(I+1,2));
end;
for I=2:N-1;
    B(I,N)=gpf(A(I-1,N)+A(I+1,N)+A(I-1,N-1)+A(I,N-1)+A(I+1,N-1));
end;
B(1,1)=gpf(A(1,2)+A(2,2)+A(2,1));
B(1,N)=gpf(A(1,N-1)+A(2,N)+A(2,N-1));
B(N,1)=gpf(A(N,2)+A(N-1,1)+A(N-1,2));
B(N,N)=gpf(A(N-1,N)+A(N-1,N-1)+A(N,N-1));
gpf2dim=B;
```

Finally, the next function receives a square matrix $A$ as an input, executes $K$ iterative steps as specified above, and displays as output a visual representation of the state reached in the end (again, this can be changed as desired). This game based on the greatest prime factor produces matrices with prime entries, but we will not need to require the seed $A$ to have prime entries (anyway, each of the subsequently produced matrices will have prime entries):

```
function [CAgpf2d] = CAgpf2d(A,K)
M=size(A,2);
for R=1:K;
    A=gpf2dim(A);
end;
for S=1:M;
    for T=1:M;
        C(S,T)=A(S,T);
    end
end
for S=1:M;
    C(M+1,S)=A(1,S);
end
for T=1:M;
    C(T,M+1)=A(T,1);
end
C(M+1,M+1)=A(1,1);
CAgpf2d=A
mesh(C)
```

As an example, let's start with the following 21-by-21 integer matrix $A = [a_{ij}]$ with $a_{ij} = i^2 + j$. The fact that it doesn't have prime entries is not particularly important, since every subsequent matrix in this GPF automaton will have prime entries.

The visual representation ("mesh" style) of the state of the automaton after 10 iterations appears as follows:

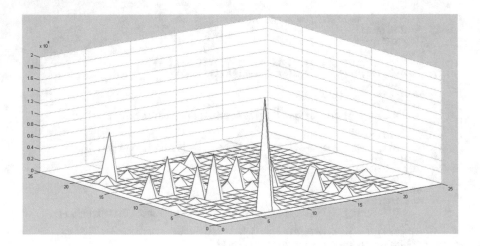

### 5.2.3   2D Conway Subprime Automata

The nonassociative 2D Conway subprime automata are defined in a similar way to the GPF ones: just replace the greatest prime factor function with the Conway subprime function, for which, again, we will need the following:

```
function lf=lf(n)
lf=min(factor(n));
end


function conway=conway(n)
if n==1
conway=1;
elseif isprime(n)==1;
    conway=n;
else
    conway=n/lf(n);
end
```

The next function represents one step in the evolution of this system. Again, the coefficients in the Moore neighborhood can be changed if needed:

```
function [ conway2dim ] = conway2dim(A)
N=size(A,2);
for I=2:N-1;
    for J=2:N-1;
        B(I,J)=conway(A(I,J+1)+4*A(I-1,J+1)+7*A(I+1,J+1)+A(I-
1,J)+A(I+1,J)+A(I-1,J-1)+A(I,J-1)+A(I+1,J-1));
    end;
end;
for J=2:N-1;
    B(1,J)=conway(A(1,J-1)+A(1,J+1)+A(2,J-1)+A(2,J)+A(2,J+1));
end;
for J=2:N-1;
    B(N,J)=conway(A(N,J-1)+A(N,J+1)+A(N-1,J-1)+A(N-1,J)+A(N-1,J+1));
end;
for I=2:N-1;
    B(I,1)=conway(A(I-1,1)+A(I+1,1)+A(I-1,2)+A(I,2)+A(I+1,2));
end;
for I=2:N-1;
    B(I,N)=conway(A(I-1,N)+A(I+1,N)+A(I-1,N-1)+A(I,N-1)+A(I+1,N-1));
end;
B(1,1)=conway(A(1,2)+A(2,2)+A(2,1));
B(1,N)=conway(A(1,N-1)+A(2,N)+A(2,N-1));
B(N,1)=conway(A(N,2)+A(N-1,1)+A(N-1,2));
B(N,N)=conway(A(N-1,N)+A(N-1,N-1)+A(N,N-1));
conway2dim=B;
```

Finally, the following function receives a square matrix as input, executes a specified number of iterative steps in this Conway subprime automaton, and provides a visual for the matrix thus obtained:

```
function [CAconway2d] = CAconway2d(A,K)
M=size(A,2);
for R=1:K;
    A=conway2dim(A);
end;
for S=1:M;
    for T=1:M;
        C(S,T)=A(S,T);
    end
end
for S=1:M;
    C(M+1,S)=A(1,S);
end
for T=1:M;
    C(T,M+1)=A(T,1);
end
C(M+1,M+1)=A(1,1);
CAconway2d=A
mesh(C)
```

For example, with the same seed matrix $A = [i^2 + j]_{ij}$ as in the case of the GPF automaton previously discussed and with the same number of iterations (10), the following output results from **CAconway2d(A,10)**:

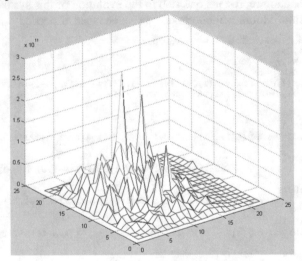

The following example has the GPF cellular automaton rule based on the following grid:

$$
\begin{array}{ccc}
1 & 6 & 3 \\
1 & \boxed{13} & 1 \\
1 & 1 & 1
\end{array}
$$

As such, it represents an example with $Z \neq 0$ that we previously announced. The basic step in the iteration is the following (also note the change in the boundary recurrence form):

```
function [ gpf2dimstep ] = gpf2dimstep(A)
%UNTITLED Summary of this function goes here
%   Detailed explanation goes here
N=size(A,2);
for I=2:N-1;
    for J=2:N-1;
        B(I,J)=gpf(13*A(I,J)+6*A(I,J+1)+A(I-1,J+1)+3*A(I+1,J+1)+A(I-
1,J)+A(I+1,J)+A(I-1,J-1)+A(I,J-1)+A(I+1,J-1));
    end;
end;
for J=2:N-1;
    B(1,J)=gpf(13*A(1,J)+A(1,J-1)+A(1,J+1)+A(2,J-1)+A(2,J)+A(2,J+1));
end;
for J=2:N-1;
    B(N,J)=gpf(13*A(N,J)+A(N,J-1)+A(N,J+1)+A(N-1,J-1)+A(N-1,J)+A(N-1,J+1));
end;
for I=2:N-1;
    B(I,1)=gpf(13*A(I,1)+A(I-1,1)+A(I+1,1)+A(I-1,2)+A(I,2)+A(I+1,2));
end;
for I=2:N-1;
    B(I,N)=gpf(13*A(I,N)+A(I-1,N)+A(I+1,N)+A(I-1,N-1)+A(I,N-1)+A(I+1,N-1));
end;
B(1,1)=gpf(13*A(1,1)+A(1,2)+A(2,2)+A(2,1));
B(1,N)=gpf(13*A(1,N)+A(1,N-1)+A(2,N)+A(2,N-1));
B(N,1)=gpf(13*A(N,1)+A(N,2)+A(N-1,1)+A(N-1,2));
B(N,N)=gpf(13*A(N,N)+A(N-1,N)+A(N-1,N-1)+A(N,N-1));
gpf2dimstep=B;
```

The following function makes *K* iterations starting from a seed configuration of the grid:

```
function [CAplusgpf2d] = CAplusgpf2d(A,K)
%UNTITLED3 Summary of this function goes here
%   Detailed explanation goes here
M=size(A,2);
for R=1:K;
    A=gpf2dimstep(A);
end;
for S=1:M;
    for T=1:M;
        C(S,T)=A(S,T);
    end
end
for S=1:M;
    C(M+1,S)=A(1,S);
end
for T=1:M;
    C(T,M+1)=A(T,1);
end
C(M+1,M+1)=A(1,1);
CAplusgpf2d=A
mesh(C)
```

The following two MATLAB functions will build a random "seed" matrix with prime entries:

```
function ithprime = ithprime(n)
x=primes(1000000);
ithprime=x(n);
```

```
function [seed] = seed( M )
for I=1:M;
    for J=1:M;
        A(I,J)=ithprime(max(1,floor(100*rand())));
    end
end
seed=A;
```

Or if we wanted a more precise seed configuration, we used something like this:

```
function [seed0] = seed0(M)
for I=1:M;
    for J=1:M;
        A(I,J)=ithprime(I^2+3*I*J+2*J^2);
    end
end
seed0=A;
```

Now let's see some results.

The images presented here are typical for the sort of "evolving complexity" displayed by the automata in which a state is updated as the greatest prime factor (or Conway's subprime function, for that matter) of a combination of the states in a certain Moore neighborhood.

**CAplusgpf2d(seed0(21),10)**

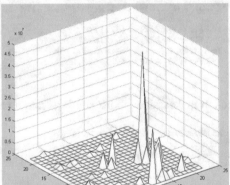

**CAplusgpf2d(seed0(21),30)**

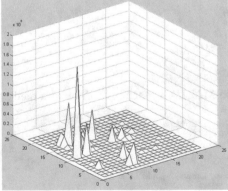

**CAplusgpf2d(seed(21),10)**

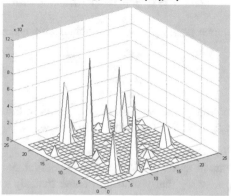

**CAplusgpf2d(seed(21),20)**

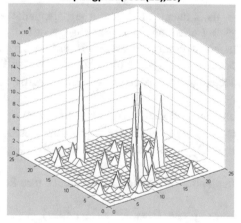

## 5.3 Walks from Greatest Prime Factor Sequences and a Mysterious Chebyshev-Like Bias

An interesting way to move forward is to look into the possibility of using the classes of sequences considered in this book (greatest prime factor sequences and Conway subprime sequences) to the generation of (pseudo)random numbers and random walks. This section will deal with GPF sequences, leaving the case of Conway subprime sequences for a further date.

Let $\{x_n\}_{n \geq 0}$ be a greatest prime factor sequence of order $d$, that is, an integer sequence satisfying

$$x_n = c_1 x_{n-1} + c_2 x_{n-2} + \cdots + c_d x_{n-d} + c_0 (n \geq d).$$

To the prime sequence $\{x_n\}_{n \geq 0}$ we can associate a binary $\pm 1$ sequence using the sequence

$$\{2 - (x_n \bmod 4)\}_{n \geq 0}.$$

The only problem with the above assignment occurs in the (fortunately typically rare) cases in which $x_n = 2$. We can solve it in two ways:

- Associate a "+1" to each $n$ for which $x_n = 2$.
- Make arrangements in terms of the recurrence coefficients $c_0, \ldots, c_d$ and/or the "seed" terms $x_0, \ldots, x_{d-1}$ that would guarantee that all terms of the sequence $\{x_n\}_{n \geq 0}$ are odd. For example, if $d$ is even and $x_0, \ldots, x_{d-1}$ are odd, we may choose $c_0, \ldots, c_d$ all odd, or $c_0 = 0$ and $c_1, \ldots, c_d$ all odd with a single exception.

In our experience, the strategy of "making arrangements for a GPF sequence with odd terms" does not effect any practical change in the general aspect of the pseudorandom walks that would be generated. It will be our preferred method.

Thus, to a GPF sequence $\{x_n\}_{n \geq 0}$ with odd terms we will associate the pseudorandom walk reflected in the sequence of partial sums of the $\pm 1$ sequence $\{2 - (x_n \bmod 4)\}_{n \geq 0}$.

Assume that the ultimate periodicity conjecture for GPF sequences, $\{x_n\}_{n \geq 0}$ will eventually enter a cycle. This is definitely a problem, because the ensuing periodicity of $\{2 - (x_n \bmod 4)\}_{n \geq 0}$ is a bad sign if we desire a good random number generator.

Consider, for example, the GPF-tribonacci sequence (see Section 3.2) satisfying

$$\begin{cases} x_n = P(x_{n-1} + x_{n-2} + x_{n-3}) \\ x_0 = x_1 = x_2 = 1 \end{cases}.$$

It turns out that the above sequence has period 212. This is fairly small, it and makes the plot of the sequence of cumulative sums of $\{2 - (x_n \bmod 4)\}_{n=0}^{1000}$ quite predictable:

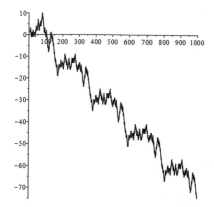

The main problem we had was to select an appropriate initial segment of the greatest prime factor sequence so that the $\pm 1$ walk based on it would have a sporting chance to be considered a fairly good model of a random number generator.

Our first attempt was to work with

$$\{2 - (x_n \bmod 4)\}_{n=0}^{\text{TENTATIVE\_START}} \text{ or } \{2 - (x_n \bmod 4)\}_{n=0}^{\text{PP}+\text{PERIOD}},$$

where the values "TENTATIVE_START," "PP" (preperiod), and "PERIOD" are found using Floyd's cycle-finding algorithm. Note that from the design of Floyd's algorithm, we have

$$\text{PP} \leq \text{TENTATIVE\_STRAT} \leq \text{PP} + \text{PERIOD}.$$

In diagrammatic terms, when we get to "TENTATIVE_START" we are in the cycle, but we did not complete the full "rho" shape yet. Typically, we used recurrences of orders $d \geq 4$ (this typically leads to longer periods and preperiods; thus the above GPF-tribonacci recurrence of order 3 does not qualify).

The result was essentially negative in the sense that the above two types of $\pm 1$ sequences cannot be considered indistinguishable from genuine 1D random walks from the point of view of the total displacement. However, we feel it is desirable to have a precise knowledge about these sequences. The computational procedure that we used to that effect will be described below.

Here is the "master" MAPLE file that we used for the exploration of these walks. First we always load the MAPLE packages that we expect to use at some point (this does not mean that all of them are used in a particular experiment; it is just our "usual header"):

```
> with(numtheory): with(ListTools): with(plots):
with(StringTools): with(RandomTools): with(Statistics):
```

The next in the list is the procedure that will eventually be used to find the preperiod through backtracking:

```
> PP:=proc(X::list, TS::integer, P::integer);
k:=TS;
while X[k]-X[k+P]=0 do
k:=k-1;
end do;
PP:=k;
end proc;
```

In the next line we will select the coefficients $a, b, c, d$ of the degree-4 GPF recurrence:

$$x_n = P(ax_{n-1} + bx_{n-2} + cx_{n-3} + dx_{n-4}).$$

Since we will be using a seed $(x_0, x_1, x_2, x_3)$ with odd prime components, we will make sure that $a + b + c + d$ is odd. We will first try relatively small (but otherwise random) values $a, b, c, d$ . In the same line we will introduce the function that makes the transition from a segment $(x_{n-4}, x_{n-3}, x_{n-2}, x_{n-1})$ to $(x_{n-3}, x_{n-2}, x_{n-1}, x_n)$:

```
> a:=rand(1..10)(); b:=rand(1..10)(); c:=rand(1..10)();
d:=(a+b+c+1) mod 2+2*rand(1..5)(); f:=(x,y,z,w)->
(y,z,w,P(a*w+b*z+c*y+d*x));
```

Next comes the actual Floyd algorithm, which produces the tentative start, the period, and the preperiod of the greatest prime factor sequence with the initial condition that $(x_0, x_1, x_2, x_3)$ are random odd primes selected from among the first 1000 primes:

$$\begin{cases} x_n = P(ax_{n-1} + bx_{n-2} + cx_{n-3} + dx_{n-4}) \\ x_0 = x0, x_1 = x1, x_2 = x2, x_3 = x3 \end{cases}$$

```
> x0:=ithprime(rand(2..1000)()); x1:=ithprime(rand(2..1000)());
x2:=ithprime(rand(2..1000)()); x3:=ithprime(rand(2..1000)());
X:=(x0,x1,x2,x3): SLOW:=f(X): FAST:=f(f(X)): k:=1: while
SLOW<>FAST do SLOW:=f(SLOW): FAST:=f(f(FAST)): k:=k+1; end do:
TENTATIVE_START:=k; X:=SLOW: SLOW:=f(X): FAST:=f(f(X)): k:=1:
while SLOW<>FAST do
SLOW:=f(SLOW): FAST:=f(f(FAST)): k:=k+1; end do: PERIOD:=k;
x(0):=x0: x(1):=x1: x(2):=x2: x(3):=x3: for r from 4 to
TENTATIVE_START+PERIOD do x(r):=P(a*x(r-1)+b*x(r-2)+c*x(r-3)
+d*x(r-4)) end do: X:=[seq(x(r),r=0..TENTATIVE_START+PERIOD)]:
PP:=PP(X,TENTATIVE_START,PERIOD);
```

Finally, we will make the conversion to $\{2 - (x_n \bmod 4)\}$ with subscripts ranging up to either TENTATIVE_START or PP + PERIOD. Note that if we choose to work with the bound TENTATIVE_START, we don't need to calculate the preperiod PP.

```
> N:=PP+PERIOD; L:=[seq(2-(x(r) mod 4),r=0..N)]:
  W:=PartialSums(L): listplot(W);
```

After a series of experiments with these order-4 GPF sequences, we saw no evidence that the ±1 sequences $\{2 - (x_n \bmod 4)\}$, based on their cumulative sums W, are accurate models of genuine random walks (however, this does not mean that such models cannot be obtained if we increase the order of the recurrence or recurrence coefficients). We don't know the reason for the relatively large displacement (see the two examples below). Since this is meant to be a lab manual, we have to record all findings and advise the reader accordingly.

*Example 5.1* Consider the recurrent sequence $\{x_n\}_{n \geq 0}$ defined by the following fourth-order GPF recurrence:

$$\begin{cases} x_n = P(x_{n-1} + 7x_{n-2} + 2x_{n-3} + 3x_{n-4}) \\ x_0 = 1307, \ x_1 = 787, \ x_2 = 6791, x_3 = 5381 \end{cases}$$

The variables TENTATIVE_START and PERIOD turned out to be 228568 and 114284, respectively. Finding TENTATIVE_START to be reasonably large, we decided to work with it as the upper bound for the sequence segment to be transformed into a ±1 walk. The plot of the partial sums of the ±1 sequence $\{2 - (x_n \bmod 4)\}_{n=0}^{228568}$ appeared as follows:

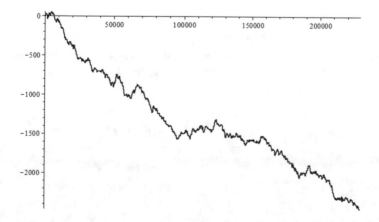

This is an obviously decreasing overall trend, with a global drop of 2457 in 228568 terms. On the other hand, the binomial standard deviation for a supposedly genuine 228568-step random walks (with $p(+) = p(-) = p = 1/2$ would be $\sqrt{N \cdot p(1-p)} = \frac{\sqrt{N}}{2} \approx 239.04$. Looking at the plot, we see that we are more than

10 such deviations away from the expected average (zero), so this cannot be considered a desirable random walk model. Also, it is known that the expected displacement of a large $\pm 1$ random walk with $N$ steps is $\frac{\sqrt{2N}}{\pi}$. In this case, this amounts to 381.4, which is significantly different from what appears above.

*Example 5.2* Consider a GPF sequence $\{x_n\}_{n \geq 0}$ defined as follows:

$$\begin{cases} x_n = P(11x_{n-1} + 3x_{n-2} + 9x_{n-3} + 4x_{n-4}) \\ x_0 = 491, x_1 = 283, x_2 = 491, x_3 = 47 \end{cases}$$

This example manifests similar traits to those in the previous sequence. See the partial sums plot below, the preperiod being 1025108, with a period of 88389. This time we decided to use PP + PERIOD for the upper limit in the range of the terms of the sequence. Again, there is a noticeable departure from genuine randomness.

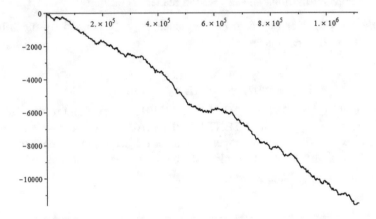

We noted the curious finding that the trends we looked into are usually (though not always, as we shall see) descending. We will discuss this in more detail later in this section, speculating on the possible causes. Still, this kind of overall global shape warrants the question asked in the following computational exploration project:

**COMPUTATIONAL EXPLORATION PROJECT CEP19.** Given the observed negative displacement leaning for "global(ish)" $\pm 1$ walks $\{2 - (x_n \bmod 4)\}_{n=0}^{\text{TENTATIVE\_START}}, \{2 - (x_n \bmod 4)\}_{n=0}^{\text{PREPERIOD} + \text{PERIOD}}$ (or $\{2 - (x_n \bmod 4)\}_{n=\text{PREPERIOD}}^{\text{PREPERIOD} + \text{PERIOD} - 1}$ as we shall see later) associated to a greatest prime factor sequence $\{x_n\}_{n \geq 0}$, find a reasonably large optimal value $T$ (which may depend on the order or the GPF recurrence) such that $\{2 - (x_n \bmod 4)\}_{n=0}^{T}$ behave, from as many respects as possible, like genuine random walks?

To us, the tentative value $T = 10000$ seemed satisfactory enough when we were working with GPF sequences of order up to 5.

Of course, the reason for this resemblance is certainly deterministic (like our construction) and may be due to a deep number-theoretic reason unknown to the author and beyond the scope of this book. We will try to suggest a possible connection with a known "bias" in the distribution of primes in the residue classes (modulo 4 in this case).

Note that we looked for this apparently consistent departure from resembling genuine randomness to GPF sequences of higher order, five or more, in which case even the execution of the cycle-finding algorithm takes a comparatively longer time. For example, consider the following prime sequence of order five:

$$\begin{cases} x_n = 3x_{n-1} + x_{n-2} + 3x_{n-3} + 4x_{n-4} + 2x_{n-5} \ (n \geq 5), \\ x_0 = 3, x_1 = 11, x_2 = 7, x_3 = 41, x_4 = 11. \end{cases}$$

This GPF sequence has a period $P = 3173647$ and preperiod $PP = 42684008$. Seen "from a distance," the plot of the cumulative sums of the sequence $\{2 - (x_n \bmod 4)\}_{n=0}^{PP + PERIOD}$ has an already familiar aspect:

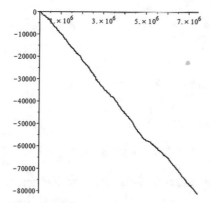

However, if we restrict ourselves to $\{2 - (x_n \bmod 4)\}_{n=0}^{10000}$, the graph looks much more like a genuine random walk, at least regarding displacement:

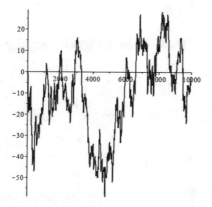

In conclusion, our unsuccessful (at least when it comes to the walks' endpoints) attempts in working with the "global" cases $\{2 - (x_n \bmod 4)\}_{n=0}^{\text{TENTATIVE\_START}}$ and $\{2 - (x_n \bmod 4)\}_{n=0}^{\text{PP} + \text{PERIOD}}$ made us switch to the idea of using a "uniform range" in the sense of working with $\pm 1$ sequences of the form

$$\{2 - (x_n \bmod 4)\}_{n=0}^{T},$$

where $T = 10000$ is a value that we found generally desirable for GPF sequences of order 4 (with the mention that it can be updated to higher values if needed. Note that the random integer generator from the excellent site https://www.random.org/integers/ is able to generate up to 10000 random numbers. For the sequence in Example 5.1, the associated $\pm 1$ sequence $\{2 - (x_n \bmod 4)\}_{n=0}^{10000}$ has partial sums displayed below:

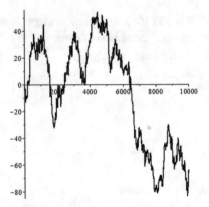

It has a downward displacement of 67, and it could be, according to the endpoint criterion, a random walk model.

For the sequence in Example 5.2, the partial sums of the corresponding $\pm 1$ sequence $\{2 - (x_n \bmod 4)\}_{n=0}^{10000}$ are plotted below. The overall (downward) displacement is 127.

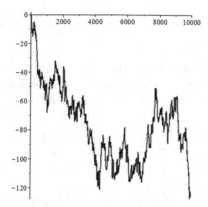

Keeping $T = 1000$, we will use a random selection of 1000 GPF sequences of order 4:

$$\begin{cases} x_n = ax_{n-1} + bx_{n-2} + cx_{n-3} + dx_{n-4} \, (4 \le n \le 10000) \\ x_0, x_1, x_2, x_3 \in \Pi \end{cases}$$

Selecting each of the 1000 sequences of the above form involves random recurrence coefficients $a, b, c, d$ as well as random seeds $(x_0, x_1, x_2, x_3)$.

To each such sequence we will associate a "relative displacement" variable represented by the quotient of the total displacement of $\{2 - (x_n \bmod 4)\}_{n=0}^{10000}$ and the square root of the sequence length, $\sqrt{10001}$. The plan is to organize those constants as components of a vector and see its histogram. Thus we will be able to visualize a fairly accurate displacement distribution.

We did this with MAPLE in the following way. We start by randomly generating 1000 GPF sequences of order 4 and length 10001, associate a $\pm 1$ string to each of them as previously specified, and enter the sum of the $\pm 1$ string as a component in the displacement vector $Z$:

```
V:=1000: for M from 1 to V do a:=rand(1..100)();
b:=rand(1..100)(); c:=rand(1..100)(); d:=((1+a+b+c) mod
2)+2*rand(1..50)(); T:=10000: x(0):=ithprime(rand(2..100)());
x(1):=ithprime(rand(2..100)()); x(2):=ithprime(rand(2..100)());
x(3):=ithprime(rand(2..100)()); for r from 4 to T do
x(r):=P(a*x(r-1)+b*x(r-2)+c*x(r-3)+d*x(r-4)) end do:
Z(M):=add(x, x in [seq(2-(x(r) mod 4),r=0..T)]; end do:
```

We finalize by producing the list of normalized displacements, together with a histogram that offers a closer look at their distribution:

```
> Y:=[seq(evalf(Z(k)/sqrt(10001)),k=1..V)]: Histogram(Y);
```

The histograms of the normalized endpoints in both 40-bin format and discrete mode with absolute frequency scale are shown below, and they indicate the distribution of the "square roots of the walk length" about the origin.

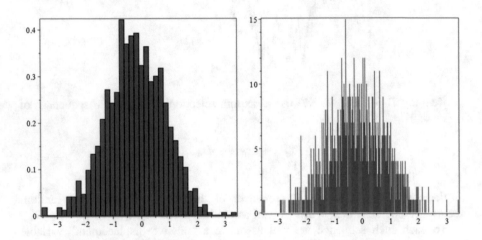

To use statistical terminology, we could say that for most walks $\{2 - (x_n \bmod 4)\}_{n=0}^{10000}$ produced in the experiment there is insufficient reason to reject the "null hypothesis" (that they behave like genuine random walks *from the point of view of the displacement*). To see this, we use a "chi-square" statistic for any particular $\pm 1$ walk (one degree of freedom, just as in the case of the coin flip experiments), which can be shown to be of the form

$$\frac{(N_+ - N_-)^2}{N} = \frac{\Delta^2}{N},$$

where $\Delta$ is the walk displacement for $Z(k)$ (for the $k$th generated walk, in the MAPLE code), and $N$ is the length of the walk (10001 as used by us). The following histogram represents the distribution of the chi-square values among the 1000 walks generated in the experiment.

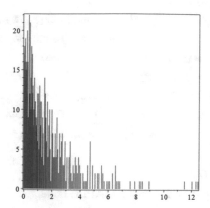

Among the 1000 GPF-based $\pm 1$ walks, 941 have the associated chi-square value less than the critical constant of 3.41, which makes those walks statistically indistinguishable (at a $p = 0.05$ cutoff) from genuine random walks from the point of view of displacement. This is somewhat encouraging. Here is a MAPLE code that can be used for the chi-square analysis:

```
V:=1000: for M from 1 to V do a:=rand(1..100)();
b:=rand(1..100)(); c:=rand(1..100)(); d:=((1+a+b+c) mod
2)+2*rand(1..50)(); T:=10000: x(0):=ithprime(rand(2..100)());
x(1):=ithprime(rand(2..100)()); x(2):=ithprime(rand(2..100)());
x(3):=ithprime(rand(2..100)()); for r from 4 to T do
x(r):=P(a*x(r-1)+b*x(r-2)+c*x(r-3)+d*x(r-4)) end do: L:=[seq(2-
(x(r) mod 4),r=0..T)]: Z(M):=sum(L[k],k=1..T+1); end do:
Y:=[seq(evalf(Z(k)^2/10001),k=1..V)];
```

The $\pm 1$ walks $\{2 - (x_n \bmod 4)\}_{n=0}^{10000}$ can be read into a "bit string" (0/1) key by switching to

$$\{-1 + (x_n \bmod 4)/2\}_{n=0}^{10000}.$$

Further, let us associate to any block of eight consecutive bits $b_1, \ldots, b_8$ in the above 0/1 sequence the following element of the unit interval $[0, 1]$:

$$\frac{b_1}{2} + \frac{b_2}{2^2} + \cdots + \frac{b_8}{2^8}.$$

Thus we can get a sequence of 1250 numbers. We can visualize their distribution in $[0, 1]$ by building a 10-bin histogram. The Maple code for this, together with explicit detail on the number of entries falling in each bin, is the following:

```
> x(0):=ithprime(rand(2..100)());
x(1):=ithprime(rand(2..100)()); x(2):=ithprime(rand(2..100)());
x(3):=ithprime(rand(2..100)()); for r from 4 to 10000 do
x(r):=P(a*x(r-1)+b*x(r-2)+c*x(r-3)+d*x(r-4)) end do:
B:=[seq((-1+x(r) mod 4)/2,r=0..10000)]:
K8:=[seq(evalf(B[8*t+1]/2+B[8*t+2]/2^2+B[8*t+3]/2^3+B[8*t+4]/2^4
+B[8*t+5]/2^5+B[8*t+6]/2^6+B[8*t+7]/2^7+B[8*t+8]/2^8),t=0..1249)
]: Histogram(K8, bincount=10, frequencyscale=absolute);
TallyInto(K8,default,bins=10);
```

We include below a typical output of the above MAPLE code.

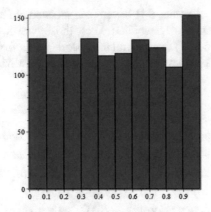

[HFloat(0.0) .. HFloat(0.099609375) = 132, HFloat(0.099609375) .. HFloat(0.19921875) = 113,

HFloat(0.19921875) .. HFloat(0.298828125) = 123, HFloat(0.298828125) .. HFloat(0.3984375) = 125,

HFloat(0.3984375) .. HFloat(0.498046875) = 124, HFloat(0.498046875) .. HFloat(0.59765625) = 116,

HFloat(0.59765625) .. HFloat(0.697265625) = 134, HFloat(0.697265625) .. HFloat(0.796875) = 118,

HFloat(0.796875) .. HFloat(0.896484375) = 113, HFloat(0.896484375) .. HFloat(0.99609375) = 152]

### 5.3.1 Self-correlations

We will now look at the self-correlation picture for $\pm 1$ sequences $\{z_n\}_{n=0}^{10000}$ of length 10001 where $z_n = 2 - (x_n \bmod 4)$ associated to an initial segment of a GPF sequence $\{x_n\}_{n \geq 0}$.

For $\pm 1$ sequences $\vec{a} = (a_0, a_1, \ldots, a_{N-1})$ the aperiodic (i.e., periodic boundary conditions are not assumed) autocorrelation coefficients are defined by

$$c_k := \sum_{j=0}^{N-k-1} a_j a_{j+k}.$$

As a concrete example, consider the recurrence of order four

$$\begin{cases} x_n = P(11x_{n-1} + 3x_{n-2} + 9x_{n-3} + 4x_{n-4}), \\ x_0 = 29, x_1 = 3, x_2 = 13, x_3 = 7. \end{cases}$$

A "step-by-step" hands-on MAPLE-based method to visualize the correlation coefficients $c_k$ with $k \neq 0$ (since obviously $c_0 = 10001$) is the following:

```
> a:=11; b:=3; c:=9; d:=4;

> f:=(x,y,z,w)->(y,z,w,P(a*w+b*z+c*y+d*x));

> x(0):=29; x(1):=3; x(2):=13; x(3):=7;

> for r from 4 to 10000 do x(r):=P(a*x(r-1)+b*x(r-2)+c*x(r-
3)+d*x(r-4)) end do:

> for T from 0 to 10000 do L[T]:=add(m, m in [seq( (2-(x(r) mod
4))*(2-(x(r+T) mod 4)),r=0..10000-T)]); end do:

> CORR:=[seq(L[T],T=1..10000)]:

> listplot(CORR);
```

The effect is the following diagram, with the maximum absolute value of a correlation coefficient with nonzero lag being 301.

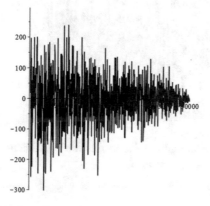

To put this in a theoretical context, we refer to the bound provided in Mercer (2006), to the effect that for every $\varepsilon > 0$ the estimate

$$\min_{\bar{a}} \left( \max_{1 \le k \le N-1} |c_k| \right) \le \sqrt{2N \ln N} + \varepsilon \sqrt{N \ln N}$$

holds for all sufficiently large $n$. For $N = 10001$ we have $\sqrt{2N \ln N} \approx 429.21$, whereas for the above particular $\pm 1$ sequence we have $\max_{1 \le k \le N-1} |c_k| = 301$.

As an alternative we have "periodic self-correlations," where the binary sequences $\{z_n\}_{n=0}^{10000}$ under consideration are set up in a "wrapped around a circle" mode: this amounts to its subscripts being considered modulo 10001. To every nonnegative integer $T$ with $0 \le T \le 10000$ we associate the following self-correlation coefficient at a distance $T$:

$$L(T) := \sum_{m=0}^{10000} z_m z_{(m+T) \bmod 10001}$$

Obviously $L(0) = 10001$. The MAPLE-based method that we used to obtain the list with components $L(T)$ for a particular GPF sequence obviously starts with getting and storing the terms of the sequence.

To get a view of periodic self-correlations, the above MAPLE code undergoes only one straightforward difference, as shown below:

```
> for T from 0 to 10000 do LP[T]:=add(m, m in [seq((2-(x(r) mod
4))*(2-(x((r+T) mod 10001) mod 4)),r=0..10000)]); end do:

> PCORR:=[seq(LP[T],T=1..10000)]: listplot(PCORR);
```

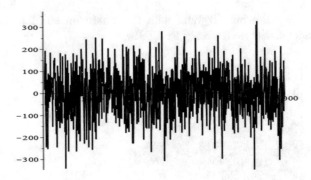

It turns out that the maximum modulus of a correlation coefficient at a nonzero distance for this particular GPF sequence is 373.

## 5.3.2   A Quasirandomness Test for the Limit Cycles of GPF Sequences

We will now discuss a quasirandomness test for cycle parts of the $\pm 1$ walks described above, in the light of the 1992 seminal paper (Chung and Graham 1992). Our action plan is as follows.

- Randomly pick random third-order GPF sequence $\{x_n\}_{n \geq 1}$ (random recurrence coefficients, random seeds in their respectively selected boxes), and execute the whole Floyd algorithm in each case, so that we know the corresponding preperiods and periods. This can be done in the same way with GPF sequences of higher order, only that the constant in terms of the running time will be higher.
- For each such sequence we will restrict to the cycle part

$$\{x_n\}_{n = \text{PREPERIOD}}^{\text{PREPERIOD} + \text{PERIOD} - 1}$$

The selection will be done by
```
LP:=[seq(x(r),r=PREPERIOD..PREPERIOD+PERIOD-1)]:
```
- Convert the above cycle part into a bit string of length "PERIOD" by assigning a 1 to those $x_n$ that are congruent to 1 modulo 4, and a 0 to the others. This can be done using the following MAPLE code:

```
identify41:=proc(n::integer);
if n mod 4 = 1 then identify41:=1;
else identify41:=0;
end if;
end proc;
```

This will be followed by

```
LP41:=[seq(identify41(LP[v]),v=1..PERIOD)]:
```

Note that the choice is arbitrary; a similar code could be written to identify the terms congruent to 3 mod 4 uniquely.
- Imagine that bit string is wrapped around the circle, with the angular position of the bit associated to $x_{\text{PREPERIOD}}$ being zero.
- Choose a random nontrivial additive character modulo the integer PERIOD and perform the sum of the character values at the positions around the circle corresponding to the 1's (it is the quasirandomness of this set is that is under investigation). One of the Chung–Graham quasirandomness criteria requires this sum to be small compared to PERIOD. To that end, we used the following MAPLE function to evaluate the (complex) absolute value of the sum:

```
H:=Q->abs(evalf((1/PERIOD)*sum(LP41[Z]*exp(2*(Z-
1)*Pi*Q*I/PERIOD),Z=1..PERIOD)));
```

Note that for each GPF sequence considered, $Q$ is randomly selected from 2 to the length of the cycle (PERIOD).

- After a number of such trials (TRIALS), each consisting of choosing a random third-order GPF sequence and calculating the corresponding exponential sum (considered relative to the cycle length, i.e., normalized by dividing it by the value of PERIOD) corresponding to the bit string associated to the limit cycle, we will create, as a final output, the histogram of the results thus obtained (organized in a list QR). Let us agree that in this "randomized approximation," the Chung–Graham quasirandomness test will be satisfied if the output histogram shows a narrow concentration of the complex moduli in the immediate right-hand neighborhood of zero (the moduli are positive).

This brings us to the main MAPLE code that allows us to look into the quasirandom character of the cycle elements congruent to 1 modulo 4, in the light of the Chung–Graham criterion EXP for quasirandomness (Chung and Graham 1992).

For example, if we want to perform an experiment with TRIALS = 100 third-order greatest prime factor sequences, with recurrence coefficients randomly selected between 1 and $K = 10$ and initial conditions randomly selected from among the first $W = 1000$ primes, we will execute the following code:

```
TRIALS:=100: for T from 1 to TRIALS do W:=1000: K:=10:
a:=rand(1..K)(): b:=rand(1..K)(): c:=rand(1..K)():
f:=(x1,x2,x3)->(x2,x3,P(c*x1+b*x2+a*x3)):
x(0):=ithprime(rand(1..W)()): x(1):=ithprime(rand(1..W)()):
x(2):=ithprime(rand(1..W)()): SEED:=(x(0),x(1),x(2)): X:=SEED:
SLOW:=f(X): FAST:=f(f(X)): k:=1: while SLOW<>FAST do
SLOW:=f(SLOW): FAST:=f(f(FAST)): k:=k+1; end do:
TENTATIVE_START:=k: Y:=SLOW: X:=SLOW: SLOW:=f(X): FAST:=f(f(X)):
k:=1: while SLOW<>FAST do SLOW:=f(SLOW): FAST:=f(f(FAST)):
k:=k+1; end do: PERIOD:=k: m:=0: U:=SEED: V:=Y: while U<>V do
m:=m+1; U:=f(U); V:=f(V) end do: PREPERIOD:=m: for r from 3 to
PREPERIOD+PERIOD do x(r):=P(a*x(r-1)+b*x(r-2)+c*x(r-3)) end do:
LP:=[seq(x(r),r=PREPERIOD..PREPERIOD+PERIOD-1)]:
LP41:=[seq(identify41(LP[v]),v=1..PERIOD)]:
Q:=rand(2..PERIOD)(): EXPTEST(T):=H(Q): end do:
QR:=[seq(EXPTEST(T),T=1..TRIALS)]: Histogram(QR);
```

Here is the output, where the histogram has an absolute frequency scale:

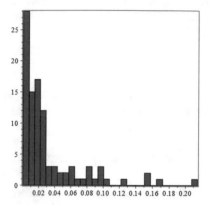

To the naked eye, it looks as though there is a high likelihood that the Chung–Graham quasirandomness EXP criterion (for the cycles of third-order greatest prime factor sequences taken modulo 4) passes. This technique (subject to further refinement by varying the constants involved) provides a judgment for a whole class of sequences and has a similar "Monte Carlo" flavor (see Appendix B).

It is motivated by the fact that if we want to apply the Chung–Graham EXP criterion even for a single GPF sequence with a high large $M$, we will have to calculate $M - 1$ exponential sums with nontrivial additive characters with $M$ terms each, which in some cases (but not always) may be costly in terms of time.

So what does this mean concretely? For a typical GPF sequence $\{x_n\}_{n \geq 1}$, if we wrap its limit cycle (assuming the ultimate periodicity conjecture that it is periodic) equally spaced around a circle, an arrangement that may be viewed as a model of the residue class ring $^\frown$PERIOD, assign 1's to cycle elements congruent to 1 modulo 4 and 0's to the others, then the set of 1's is a "quasirandom subset of $^\frown$PERIOD" according to the EXP criterion. See Chung and Graham (1992), where a number of criteria all equivalent to EXP are discussed.

Implicitly, this means we typically expect the sum graph $G$ with vertex set $^\frown$PERIOD, where two vertices are joined by an edge if their sum in $^\frown$PERIOD corresponds to a "1" (that is, indicates the position of a limit cycle element of $\{x_n\}_{n \geq 1}$ that is congruent to 1 mod 4), to be quasirandom, i.e., all $\binom{s}{2}$ labeled unoriented graphs with $s$ vertices are supposed to appear in asymptotically uniform proportion as subgraphs of $G$ (also see the related article (Chung et al. 1989) on quasirandom graphs).

### 5.3.3   The Curious Negative-Leaning Trend

We return now to the tendency for "negative displacement" that we found when we began to analyze the random $\pm 1$ walks $\{2 - (x_n \bmod 4)\}_n$ associated to greatest prime factor sequences $\{x_n\}_{n \geq 0}$. We can rewrite the previous MAPLE code for testing the quasirandom character of the distribution of cycle elements in residue classes modulo 4, so that it deals with the sums over the cycle elements

$$\Sigma_{\text{CYCLE}}\left(\{x_n\}_{n \geq 0}\right) = \sum_{n = \text{PREPERIOD}}^{\text{PREPERIOD} + \text{PERIOD} - 1} (2 - (x_n \bmod 4)).$$

For the above sums, we will not bother with the (otherwise minor) concern of a 2 occurring in the sequence (thus accepting a "zero step" in the walk). The changes in the code involve the sequence

```
> LPSIGN:=[seq(2-(x(r) mod 4),r=PREPERIOD..PREPERIOD+PERIOD-1)]:
```

Then, with $T$ representing the sampling variable (ranging, as before, from 1 to the designated value TRIALS) the above sum over the cycle elements will be

```
> SUMTEST(T):=add(x, x in LPSIGN):
```

The above entries, organized in a list SUMCYCLE, will be visualized in a histogram format. With bounds TRIALS, $W$, $K$ as in the case of the quasirandomness test, the adapted MAPLE code becomes

```
> TRIALS:=100: for T from 1 to TRIALS do W:=1000: K:=10:
a:=rand(1..K)(): b:=rand(1..K)(): c:=rand(1..K)():
f:=(x1,x2,x3)->(x2,x3,P(c*x1+b*x2+a*x3)):
x(0):=ithprime(rand(1..W)()): x(1):=ithprime(rand(1..W)()):
x(2):=ithprime(rand(1..W)()): SEED:=(x(0),x(1),x(2)): X:=SEED:
SLOW:=f(X): FAST:=f(f(X)): k:=1: while SLOW<>FAST do
SLOW:=f(SLOW): FAST:=f(f(FAST)): k:=k+1; end do:
TENTATIVE_START:=k: Y:=SLOW: X:=SLOW: SLOW:=f(X): FAST:=f(f(X)):
k:=1: while SLOW<>FAST do SLOW:=f(SLOW): FAST:=f(f(FAST)):
k:=k+1; end do: PERIOD:=k: m:=0: U:=SEED: V:=Y: while U<>V do
m:=m+1; U:=f(U); V:=f(V) end do: PREPERIOD:=m: for r from 3 to
PREPERIOD+PERIOD do x(r):=P(a*x(r-1)+b*x(r-2)+c*x(r-3)) end do:
LP:=[seq(x(r),r=PREPERIOD..PREPERIOD+PERIOD-1)]: LPSIGN:=[seq(2-
(x(r) mod 4),r=PREPERIOD..PREPERIOD+PERIOD-1)]:
SUMTEST(T):=add(x, x in LPSIGN): end do:
SUMCYCLE:=[seq(SUMTEST(T),T=1..TRIALS)]; Histogram(SUMCYCLE);
```

So that our eyes will not fool us, we also want to see SUMCYCLE explicitly. Here are the results in a typical run (absolute frequency scale for the histogram). In the listing, the nonnegative values are emphasized. The results confirm the observed "negative leaning." However, we also discover sequences with positive walk displacement over the cycle interval.

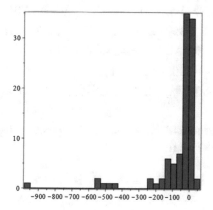

$$SUMCYCLE = [-513, -17, -15, -11, -482, -17, 6, -1, -114, -224, -41,$$
$$-100, -20, -537, 26, -158, -34, 0, -6, -4, -20, -21, -13, -6,$$
$$-1, -12, 8, -2, -54, -22, -18, -20, -8, 4, -141, 1, 2, -13, 0,$$
$$-176, -983, -10, 10, -3, -27, -67, -12, -42, 5, -47, -122, 2,$$
$$-11, -38, 6, 29, 16, -2, -64, 4, -7, -9, 9, -94, -121, -204, -6,$$
$$-13, 31, -34, -58, -1, -114, -15, -24, -5, -109, -223, 3, -84,$$
$$-31, -120, -79, -451, 64, -39, 13, 6, 5, 13, 6, -3, -4, -19, -34, 0,$$
$$-14, -548, 2, -27]$$

Let us adapt the MAPLE code and apply the same displacement check to the sums previously considered,

$$\Sigma_{TS}\left(\{x_n\}_{n \geq 0}\right) := \sum_{n=0}^{\text{TENTATIVE\_START}} (2 - (x_n \bmod 4))$$

and

$$\Sigma_{PP+P}\left(\{x_n\}_{n \geq 0}\right) := \sum_{n=0}^{\text{PREPERIOD + PERIOD}} (2 - (x_n \bmod 4)).$$

For $\Sigma_{TS}(\{x_n\}_{n \geq 0})$ a typical result in list format concluding 100 random trials is as follows:

$$[-37, -252, -19, -131, -75, -360, -1, -249, -57, -47, -17, -91, -78,$$
$$-106, -59, -34, 9, -145, -16, -22, 7, -13, -7, -197, -31, -47, -37, -52,$$
$$-765, -507, -27, -5, -27, -38, -190, -41, -6, -2, -289, 27, -201, -86, -71,$$
$$-87, -62, -84, -383, -34, -56, -35, -95, -453, -501, -95, 20, -133, -128,$$
$$-367, -130, -213, 121, -42, -302, -51, -5, -24, -6, -73, -10, -108, -523,$$
$$-113, -5, -377, -53, -15, -8, -590, -12, -111, -18, -13, -11, -215, -12,$$
$$-136, -532, -397, -11, -45, -14, 43, -325, -245, -92, -17, -18, 4, -11, -79]$$

We will have to look carefully to find the few cases with positive displacements. For $\Sigma_{PP+P}(\{x_n\}_{n \geq 0})$ a typical result is

$$[-128, -20, -56, -224, -107, -80, 32, -16, -67, -5, -363, -197,$$
$$-694, -72, -459, -194, -6, -73, -25, -46, -10, -46, -569, -689,$$
$$-46, -273, -56, -18, -201, -26, 26, -345, -21, -60, -2, -20, -175,$$
$$-93, -101, -484, -15, -810, 18, -17, -90, -53, -12, -39, -277, -231,$$
$$-994, -96, -228, -105, -152, -116, -97, -76, -135, -48, -328, 2, -249,$$
$$-3, -73, -89, -587, -193, -109, -50, -173, -530, -75, -31, -183, -6, -38,$$
$$-118, -7, 0, -27, -109, -127, 1, -88, -82, -792, -341, -23, 6, -755, -86, -263,$$
$$-44, -1, -77, -1, -49, -453, -343]$$

In conclusion, we can say that there are, though relatively very few, cases of positive displacements over the above ranges.

What is the cause of this mysterious negative trend?

We can only speculate a connection with the so called "Chebyshev bias" (Granville and Martin 2006) to the effect that there are more primes congruent to 3 modulo 4 than primes congruent to 1 mod 4.

For a start, the following is a plot representing the cumulative sums of the $\pm 1$ sequence $\{2 - (p_n \bmod 4)\}_{n=2}^{500000}$, where $p_n$ is the $n$th prime.

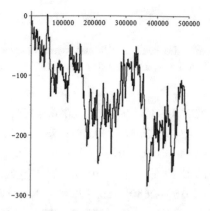

The pattern 3 mod 4 > 1 mod 4 holds most of the time, with infinitely many exceptions, though it is known (Littlewood 1914) that there are arbitrarily large values of $x$ for which $\#\{p \le x | p \equiv 1 \bmod 4\} > \#\{p \le x | p \equiv 3 \bmod 4\}$ (however, the pattern 1 mod 4 > 3 mod 4, when achieved, is short-lived. See Granville and Martin (2006) and the references therein, where the following interesting conjecture is mentioned:

**CONJECTURE** (Knapowski and Turán 1962): As $X \to \infty$, the proportion of numbers $x \le X$ with the property that $\#\{p \le x | p \equiv 3 \bmod 4\} > \#\{p \le x | p \equiv 1 \bmod 4\}$ approaches 1.

Note that Dirichlet's theorem guarantees that

$$\lim_{x \to \infty} \frac{\#\{p \le x | p \equiv 3 \bmod 4\}}{\#\{p \le x | p \equiv 1 \bmod 4\}} = 1.$$

Now, do the primes occurring in our greatest prime factor sequences satisfy the Chebyshev bias? Is the Chebyshev bias the very nature of the overall negative displacement leaning observed? That is an interesting question, because if such is the case, we generally expect more minuses in the associated pseudorandom walks, since $2 - (x_n \bmod 4) = -1 \Leftrightarrow x_n \equiv 3 \bmod 4$.

We extended this sort of exploration to $\pm 1$ walks associated to GPF sequences of higher orders (although the computation will require more time due to the increased lengths of the cycles). A similar batch of 100 trials involving the limit cycle parts

$$\{x_n\}_{n = \text{PREPERIOD}}^{\text{PREPERIOD} + \text{PERIOD} - 1}$$

of the $\pm 1$ walks with steps $2 - (x_n \bmod 4)$ associated to GPF sequences of order 4 with the coefficients $a, b, c, d$ of the recurrence

$$x_n = P(ax_{n-1} + bx_{n-2} + cx_{n-3} + dx_{n-4})$$

randomly chosen from 1 to 10 and the seed components randomly chosen from among the first 1000 primes produced as output the following list of 100 limit cycle sums:

$[-303, -1704, -81, -1935, -15160, -1500, -169, -8080, -156, -5029,$
$\quad - 612, -3731, -7324, -1239, -345, -365, -2548, -7578, -2123, -851,$
$\quad - 10273, -503, -6, -7198, -144, -389, -1616, -15330, -645, -2674,$
$\quad - 1318, -49, -12893, -11373, 5, -66, -673, -3079, -2057, -233, -1207,$
$\quad - 120, -108, -4259, -368, -7515, -2650, -63, -327, -53, -3347, -454,$
$\quad - 1250, -76, -901, -48, -5246, -20, -6953, -2703, -557, -2191, -4779,$
$\quad - 7892, -5055, -127, -9, -2, -16, -12791, -24, -10685, -9481, -771,$
$\quad - 137, -332, 196, -3459, -69, -1795, -1203, -60, -15, -480, -83, -189,$
$\quad - 98, -74, -319, -26, -6585, -116, -632, -3910, -19, -2690, -10709,$
$\quad - 602, -2104, -305]$

For better visualization, here is the corresponding histogram (absolute frequency scale):

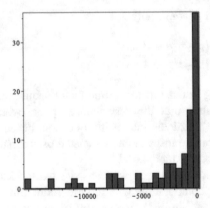

Out of 100 sequences, only two (emphasized above) have a positive displacement for the $\pm 1$ walks associated to their cycle parts. A good thing is that we discovered such sequences with positive displacements, even if they are rare.

Obviously, a negative displacement for the cycle part of an ultimately periodic sequence indicates a quasilinear long-term evolution for $\{2 - (x_n \bmod 4)\}_{n \geq 0}$.

### 5.3.4 Limitations and Opportunities

Still, we have to express some caution and mention some limitations regarding this mysterious phenomenon that suggests an analogy with the Chebyshev bias.

First of all, much more computational data should be acquired that would involve greatest prime factor sequences with limit cycles having larger terms, so that we see that the pattern of negative displacement still holds. Some sort of "high-end" computing is necessary.

Secondly, the (simplistic) hypothesis of "random sequential generation," i.e., that somehow a greatest prime factor sequence functions like a recursive device performing a sequential random sampling of primes from a certain interval, so that in each new calculation the probability of getting a term congruent to 3 modulo 4 is slightly larger than the probability of getting one that is congruent to 1 modulo 3, does not seem to be accurate. Indeed, it appears that the negative displacements in the $\pm 1$ walks $\{2 - (x_n \bmod 4)\}$ are significantly larger and far more common than they would have been under the "random sequential generation" hypothesis.

So it's not that simple. Which is a good thing, and we only hope that this is an opportunity to take an advanced look into this phenomenon from the direction of analytic number theory (beyond the scope of this book, though). For now, we just enjoy the mystery, of course with our eye on an (open-ended for now) computational project.

> **COMPUTATIONAL EXPLORATION PROJECT CEP 20.** What is the source of the observed Chebyshev-like bias (negative displacement leaning)? Is there any relation between the bias in distribution modulo 4 of terms of GPF sequences and the classical Chebyshev bias? Gain more computational evidence in support of such a bias and try to confirm or contradict the initial assessment that the discovered GPF appears to be stronger than the classical Chebyshev bias. In relation to CEP19, is there any reasonably large vale of $T$ such that the $\pm 1$ walks $\{2 - (x_n \bmod 4)\}_{n=0}^{T}$ are more likely to escape the aforementioned bias?

### 5.3.5 2D and 3D Walks

Having discussed $\pm 1$ walks with steps $\{2 - (x_n \bmod 4)\}$ associated to GPF sequences $\{x_n\}$, let us also investigate related two- and three-dimensional walks.

For 2D walks, one possibility is to consider adjacent even- and odd-ranked terms of the $\pm 1$ sequence with terms $\varepsilon_n : 2 - (x_n \bmod 4)$ as the components of a step $(u_n, v_n)$ in the 2D walk, respectively:

$$u_n := \varepsilon_{2n},$$
$$v_n := \varepsilon_{2n+1}.$$

The steps in this kind of 2D walk are $(1,1)$, $(1,-1)$, $(-1,1)$, or $(-1,-1)$. As an example, let us consider the fourth-order GPF sequence defined as follows:

$$\begin{cases} x_n = P(12x_{n-1} + 17x_{n-2} + 7x_{n-2} + 3x_{n-4}) \\ x_0 = 211, x_1 = 419, x_2 = 787, x_3 = 997 \end{cases}$$

The output of Floyd's algorithm is

TENTATIVE_START = PERIOD = 752812,

PREPERIOD = 181499.

We would like to visualize the period-related sequence

$$\{2 - (x_n \bmod 4)\}_{n=\text{PREPERIOD}}^{\text{PREPERIOD}+\text{PERIOD}-1}$$

by splitting its terms into even-ranked and odd-ranked terms as shown above. The MAPLE code that we used to do this involves the ***pointplot (U,V, style = line)*** function, which receives two input coordinate lists, $U$ for the $x$ coordinates and $V$ for the $y$ coordinates. The lists $U$ and $V$ are defined as the partial sums of even- and the odd-indexed terms of the walk $\{2 - (x_n \bmod 4)\}$, respectively, so that in the 2D plot the pair of sums

$$(\varepsilon_0 + \varepsilon_2 + \cdots + \varepsilon_{2r-2}, \varepsilon_1 + \varepsilon_3 + \cdots + \varepsilon_{2r-1})$$

is connected by a line segment to the subsequent pair

$$\varepsilon_0 + \varepsilon_2 + \cdots + \varepsilon_{2r-2} + \varepsilon_{2r}, \varepsilon_1 + \varepsilon_3 + \cdots + \varepsilon_{2r-1} + \varepsilon_{2r+1}$$

```
> for r from 4 to PREPERIOD+PERIOD do x(r):=P(a*x(r-1)+b*x(r-
2)+c*x(r-3)+d*x(r-4)) end do: X:=[seq(x(PREPERIOD+2*r-
2),r=1..PERIOD/2)]: Y:=[seq(x(PREPERIOD+2*r-1),r=1..PERIOD/2)]:
u(0):=0: for r from 1 to PERIOD/2 do u(r):=u(r-1)+2-(X[r] mod 4)
end do: U:=[seq(u(r),r=1..PERIOD/2)]: v(0):=0: for r from 1 to
PERIOD/2 do v(r):=v(r-1)+2-(Y[r] mod 4) end do:
V:=[seq(v(r),r=1..PERIOD/2)]: pointplot(U,V,style=line);
```

The output (shown below) may be considered a two-dimensional visualization of the limit cycle. Its position in the third quadrant reflect the "Chebyshev-like bias" that we discussed for the one-dimensional associated $\pm 1$ walks.

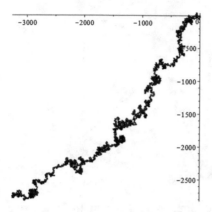

Since we are here, for the same sequence let's see a plot of the three-dimensional walk corresponding to its limit cycle, obtained using a MAPLE code involving the function **pointplot3d** *(U,V,W, style = line)*, which receives three input coordinate lists, $U$ for the $x$ coordinates, $V$ for the $y$ coordinates, and $W$ for the $z$ coordinates, so that in the 3D, plot the point

$$\left(\varepsilon_0 + \varepsilon_3 + \cdots + \varepsilon_{3r-3}, \varepsilon_1 + \varepsilon_4 + \cdots + \varepsilon_{3r-2}, \varepsilon_2 + \varepsilon_5 + \cdots + \varepsilon_{3r-1}\right)$$

is connected to the subsequent point

$$\left(\varepsilon_0 + \varepsilon_3 + \cdots + \varepsilon_{3r-3} + \varepsilon_{3r}, \varepsilon_1 + \varepsilon_4 + \cdots + \varepsilon_{3r-2} + \varepsilon_{3r+1}, \varepsilon_2 + \varepsilon_5 + \cdots + \varepsilon_{3r-1} + \varepsilon_{3r+2}\right),$$

for $r = 1, 2, 3, \ldots$. The corresponding MAPLE code follows:

```
>  M:=floor(PERIOD/3): X:=[seq(x(PREPERIOD+3*r-3),r=1..M)]:
Y:=[seq(x(PREPERIOD+3*r-2),r=1..M)]:
Z:=[seq(x(PREPERIOD+3*r),r=1..M)]: u(0):=0: for r from 1 to M do
u(r):=u(r-1)+2-(X[r] mod 4) end do: U:=[seq(u(r),r=1..M)]:
v(0):=0: for r from 1 to M do v(r):=v(r-1)+2-(Y[r] mod 4) end
do: V:=[seq(v(r),r=1..M)]:w(0):=0: for r from 1 to M do
w(r):=w(r-1)+2-(Z[r] mod 4) end do: W:=[seq(w(r),r=1..M)]:
pointplot3d(U,V,W,style=line);
```

The 3D output is shown below. Again, the mysterious "Chebyshev-like bias" is visible.

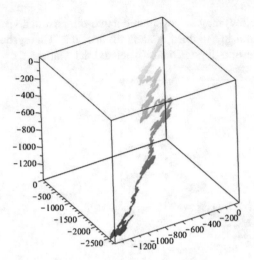

Let's continue with walks of limited range with the walk step components depending on the terms of the $\pm 1$ sequence $\{2 - (x_n \bmod 4)\}_{n=0}^{T}$, as shown below. We will work with fourth-order GPF sequences $\{x_n\}_n$ satisfying a recurrence $x_n = P(ax_{n-1} + bx_{n-2} + cx_{n-3} + dx_{n-4})$ with coefficients $a, b, c, d$ and initial conditions $x_0, x_1, x_2, x_3$ larger than in the previous examples.

For 2D walks we will adopt two plotting strategies, corresponding to two different styles of walk design.

The first one is similar to that previously used, with the points in the walk being

$$(\varepsilon_1 + \varepsilon_3 + \cdots + \varepsilon_{2r-1} + \varepsilon_2 + \varepsilon_4 + \cdots + \varepsilon_{2r}) \text{ for } r = 1, 2, \ldots, T/2.$$

Here we ignore the term indexed by 0 as well as the possible terms equal to 2 (worth the risk, since if there are any, they are very few).

Let's call this first method, *congruential* style.

In the second walk, instead of an even/odd approach, we will split the $\pm 1$ sequence $\{2 - (x_n \bmod 4)\}_{n=1}^{T}$ into two "halves" $\{2 - (x_n \bmod 4)\}_{n=1}^{T/2}$ and $\{2 - (x_n \bmod 4)\}_{n=(T/2)+1}^{T}$, so that the points in the walk will this time be of the form

$$\left(\varepsilon_1 + \varepsilon_2 + \cdots + \varepsilon_r, \varepsilon_{(T/2)+1} + \varepsilon_{(T/2)+2} + \cdots + \varepsilon_{(T/2)+r}\right), 1 \leq r \leq T/2.$$

Let's call this second method, *block* style.

We attempted to incorporate the above styles in the same block of MAPLE code so that we can compare the outputs side by side. For example, let us consider the GPF sequence $\{x_n\}_{n=0}^{10000}$ defined by the fourth-order recurrence

$$\begin{cases} x_n = P(3742x_{n-1} + 6818x_{n-2} + 3190x_{n-2} + 7627x_{n-4}) \\ x_0 = 942811, \ x_1 = 491213, \ x_2 = 65677, \ x_3 = 584509 \end{cases}$$

The output for the congruential walk style is the following 5000-step walk:

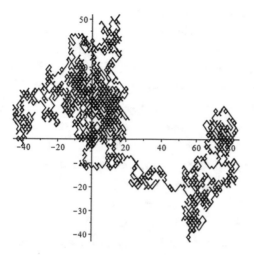

The output for the block walk style is the following 5000-step walk:

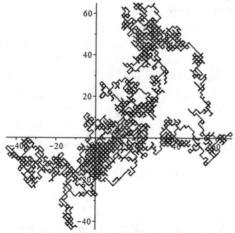

Here is the MAPLE code we used to generate pairs of congruential- and block-style walks corresponding to a randomly chosen GPF sequence of order 4:

```
> T:=10000: a:=rand(1..10000)(); b:=rand(1..10000)();
c:=rand(1..10000)(); d:=rand(1..10000)();
x(0):=nextprime(rand(1..1000000)());
x(1):=nextprime(rand(1..1000000)());
x(2):=nextprime(rand(1..1000000)());
x(3):=nextprime(rand(1..1000000)()); for r from 4 to T do x(r):=
P(a*x(r-1)+b*x(r-2)+c*x(r-3)+d*x(r-4)) end do: X:=[seq(x(2*r-
1),r=1..T/2)]: Y:=[seq(x(2*r),r=1..T/2)]: u(0):=0: for r from 1
to T/2 do u(r):=u(r-1)+2-(X[r] mod 4) end do:
U:=[seq(u(r),r=1..T/2)]: v(0):=0: for r from 1 to T/2 do
v(r):=v(r-1)+2-(Y[r] mod 4) end do: V:=[seq(v(r),r=1..T/2)]:
pointplot(U,V,style=line); X1:=[seq(x(r),r=1..T/2)]:
Y1:=[seq(x(r),r=(T/2)+1..T)]: u1(0):=0: for r from 1 to T/2 do
u1(r):=u1(r-1)+2-(X1[r] mod 4) end do:
U1:=[seq(u1(r),r=1..T/2)]: v1(0):=0: for r from 1 to T/2 do
v1(r):=v1(r-1)+2-(Y1[r] mod 4) end do:
V1:=[seq(v1(r),r=1..T/2)]: pointplot(U1,V1,style=line);
```

For 3D walks corresponding to GPF sequences $\{x_n\}_{n=0}^{T}$, we can also define, in the same way, "congruential" and "block" walks, with the congruential ones based on the splitting of the subscripts into congruence classes modulo 3, while the block ones are based on the splitting of the subscripts from 1 to $T$ into three equal blocks (as before, we will ignore the term indexed by 0 as well as possible terms equal to 2).

As an example, we will take the sequence $\{x_n\}_{n=0}^{18000}$ defined as follows:

$$\begin{cases} x_n = P(6925x_{n-1} + 578x_{n-2} + 4326x_{n-2} + 3118x_{n-4}) \\ x_0 = 662864, x_1 = 430739, x_2 = 440893, x_3 = 519031 \end{cases}$$

The 6000-step congruential-style 3D walk associated to the above GPF sequence is

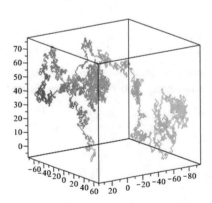

while the block style looks like this:

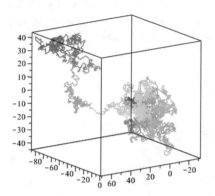

The MAPLE code used to produce the congruential–block pair for a randomly chosen GPF sequence of order 4 is included below. The parameters can be changed at will.

```
> T:=18000: a:=rand(1..10000)(); b:=rand(1..10000)();
c:=rand(1..10000)(); d:=rand(1..10000)();
x(0):=nextprime(rand(1..1000000)());
x(1):=nextprime(rand(1..1000000)());
x(2):=nextprime(rand(1..1000000)());
x(3):=nextprime(rand(1..1000000)()); for r from 4 to T do x(r):=
P(a*x(r-1)+b*x(r-2)+c*x(r-3)+d*x(r-4)) end do:
X:=[seq(x(3*r),r=1..T/3)]: Y:=[seq(x(3*r-1),r=1..T/3)]:
Z:=[seq(x(3*r-2),r=1..T/3)]: u(0):=0: for r from 1 to T/3 do
u(r):=u(r-1)+2-(X[r] mod 4) end do: U:=[seq(u(r),r=1..T/3)]:
v(0):=0: for r from 1 to T/3 do v(r):=v(r-1)+2-(Y[r] mod 4) end
do: V:=[seq(v(r),r=1..T/3)]: w(0):=0: for r from 1 to T/3 do
w(r):=w(r-1)+2-(Z[r] mod 4) end do: W:=[seq(w(r),r=1..T/3)]:
pointplot3d(U,V,W,style=line); X1:=[seq(x(r),r=1..T/3)]:
Y1:=[seq(x(r),r=T/3+1..2*T/3)]: Z1:=[seq(x(r),r=2*T/3+1..T)]:
u1(0):=0: for r from 1 to T/3 do u1(r):=u1(r-1)+2-(X1[r] mod 4)
end do: U1:=[seq(u1(r),r=1..T/3)]: v1(0):=0: for r from 1 to T/3
do v1(r):=v1(r-1)+2-(Y1[r] mod 4) end do:
V1:=[seq(v1(r),r=1..T/3)]: w1(0):=0: for r from 1 to T/3 do
w1(r):=w1(r-1)+2-(Z1[r] mod 4) end do:
W1:=[seq(w1(r),r=1..T/3)]: pointplot3d(U1,V1,W1,style=line);
```

We noticed that a shadow of the "Chebyshev-like bias" can be observed for a significant number of such fixed-range walks with a sufficiently large number of terms. For a small value of $T$, more or less anything could happen, so in general, no bias is apparent. Here is a view of such a (congruential style) walk with 400 steps:

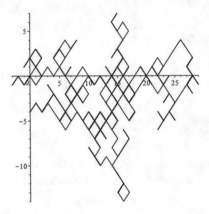

But as $T$ increases, again in reference to CEP20, could a "right balance" for the number of terms $T$ be found such that such bias is (for most cases) unnoticeable? Does the bias attenuate to some extent if we increase the size of the recurrence coefficients and the seed components? (To us, it seems that it does, although more experiments are necessary).

### 5.3.6   WYSIWYG (Well, Almost...)

Let us try another way to visualize GPF sequences. Instead of going through the $\pm 1$ sequence $\{2 - (x_n \bmod 4)\}_n$, let us try to adopt a "what you see is what you get" (WYSIWYG) style and visualize the prime sequences $\{x_n\}_{n \geq 0}$ themselves. In general, however, this poses some problems, due to the extreme differences in the sizes of the $x_n$'s. To smooth out the terms, the next best thing we could do was to find ways of visualizing an appropriate segment of the sequence $\{\ln(x_n)\}_{n=1}^{T}$ of logarithms. As done before, we will plot $\{\ln(x_n)\}_n$ in the congruential style and in the block style. We will present some typical shapes for these walks. Note some differences, as follows:

- In the congruential style, the points of the walk that will be plotted are $(\ln(x_{2r-1}), \ln(x_{2r}))$, $r = 1, 2, \ldots T/2$.
- In the block style, the points of the walk that will be plotted are $\big(\ln(x_r), \ln(x_{(T/2)+r})\big)$, $r = 1, 2, \ldots T/2$.

For the GPF sequence $\{x_n\}_{n=0}^{6000}$, we have

$$\begin{cases} x_n = P(81x_{n-1} + 63x_{n-2} + 36x_{n-2} + 100x_{n-4}) \\ x_0 = 720481, x_1 = 892079, x_2 = 745103, x_3 = 20231 \end{cases}$$

*The congruential and block walks* (respectively) are displayed below.

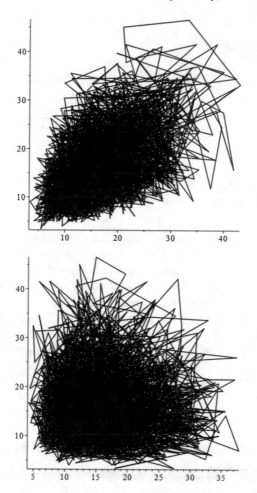

In three dimensions, we obtain a structure analogous to those above: beautiful hedgehogs! A typical output, corresponding to the GPF sequence $\{x_n\}_{n=0}^{12000}$ satisfying the recurrence

$$\begin{cases} x_n = P(61x_{n-1} + 61x_{n-2} + 31x_{n-2} + 38x_{n-4}) \\ x_0 = 670919, x_1 = 389507, x_2 = 768343, x_3 = 22787 \end{cases}$$

is shown below (this one is in congruential style; the block style output is left for the reader):

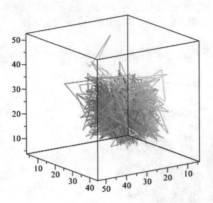

## 5.4    Linear Complexity of Bitstreams Derived from GPF Sequences

In the previous section we explored the $\pm 1$ sequences and related walks based on greatest prime factor sequences. The fact that they are marred by a strange Chebyshev-like bias probably does not make them a good source of random number generators for simulation-related purposes (or at least not without some additional processing). On the other hand, their behavior over the limit cycles appears to satisfy the "EXP" criterion for quasirandomness. These two properties are not necessarily at odds, since there are quasirandom subsets of rings $\mathbb{Z}_N$ with asymptotic densities larger or smaller than 0.5. For example, by analogy with the quasirandom subset of $F_p$ consisting of the $(p-1)/2$ nonzero perfect squares modulo $p$, if $p \equiv 1 \pmod{N}$ is a prime, $\psi$ a nontrivial multiplicative character modulo $p$ of order $N$, $\omega$ a primitive root of unity of order $N$, and $0 \leq k \leq N - 1$, then the set $\{t \in F_p | \psi(t) = \omega^k\}$ is a quasirandom subset of $F_p$ of density approaching $1/k$ as $p \to \infty$.

In this section we will analyze bitstreams (0/1 sequences) associated with greatest prime factor sequences

$$\begin{cases} x_n = P(a_1 x_{n-1} + a_2 x_{n-2} + \cdots + a_d x_{n-d}) \text{ for } n \geq d \\ x_0, x_1, \ldots, x_{d-1} \in \Pi \end{cases}$$

by associating a 1 to every sequence term $x_n$ satisfying $x_n = 3 \pmod 4$, and a 0 to every sequence term $x_n$ satisfying $x_n = 1 \pmod 4$ or $x_n = 2$ (although the occurrence of $x_n = 2$ is very rare, we will have to account for it). Converting the primes in the GPF sequence $\{x_n\}_{n \geq 0}$ to bits can be done quickly with a simple MAPLE code:

```
gpf2stream:=proc(x::integer);
if (x mod 4 = 1) or (x=2) then K:=0;
else K:=1;
end if;
end proc;
```

For example, if we use the notation $u_n = \text{gpf2stream}(x_n)$, the list of the first 101 terms $\{u_n\}_{n=0}^{100}$ in the bitstream corresponding to the GPF sequence $\{x_n\}_{n \geq 0}$ defined by the recurrence

$$\begin{cases} x_n = P(7x_{n-1} + 13x_{n-2} + 19x_{n-3} + 3x_{n-4}) \\ x_0 = 137, x_1 = 89, x_2 = 997, x_3 = 11 \end{cases}$$

is the following:

$$\begin{aligned} s := [&0,0,0,1,0,1,0,0,1,0,0,1,1,0,1,0,0,0,0,1,1,0,1,1,1,1,1, \\ &0,0,1,1,0,1,0,1,0,0,0,1,0,0,0,1,0,0,1,1,0,0,1,1,0,1,0,1,0,1, \\ &0,1,0,0,0,1,0,0,0,0,1,0,0,1,0,0,0,0,1,1,1,0,0,1,1,1,1,1,0,1, \\ &0,0,0,0,0,0,0,1,1,1,0,1] \end{aligned}$$

An important assessment measure of the suitability of a bitstream sequence $\{u_n\}$ (with terms viewed as elements of $\mathbb{Z}/2\mathbb{Z}$) for cryptographic applications is its linear complexity, that is, the degree of the smallest linear recurrence over $\mathbb{Z}/2\mathbb{Z}$ that generates the sequence (in other words, the shortest linear feedback shift register generating it). A small linear complexity is undesirable, since it means a "predictable" sequence that is insecure as a stream cipher, due to the fact that the relatively simple key to the sequence generation can be easily recovered from a fairly limited number of terms.

In what follows we will try to provide some arguments and some computational evidence supporting the hypothesis that the bitstreams produced as stated above from GPF sequences are generally acceptable as stream ciphers from the point of view of linear complexity.

The main reasons can be summarized as follows:

- The periods of the bitstreams $\{u_n\}_{n \geq 0}$ are expected to be large, especially if they derive from sequences associated to GPF sequences of higher degrees.
- For such bitstreams (or significant segments of them), the Berlekamp–Massey algorithm indicates a high linear complexity.
- Nonlinearity: if a bitstream $\{u_n\}_{n \geq 0}$ is derived from the GPF sequence $\{x_n\}_{n \geq 0}$, the prime sequence $\{x_n\}_{n \geq 0}$ is hidden from the generic observer, while the bitstream $\{u_n\}_{n \geq 0}$ is not itself produced by a linear recurrence over $\mathbb{Z}/2\mathbb{Z}$.

The following MAPLE procedure for the Berlekamp–Massey algorithm is from the website planetmath.org (http://planetmath.org/mapleimplementationofberle kampmasseyalgorithm). It calculates the polynomial generator pertaining to the first $2N$ terms of the input sequence $s$ with at least $2N$ terms (the value $P$ will be taken to be 2, since we are working with bitstreams $\{u_n\}$ over $\mathbb{Z}/2\mathbb{Z}$).

```
BM:= proc(s, N, P, x)
   local C,B,T,L,k,i,n,d,b,safemod;
   ASSERT(nops(s)=2*N);
   safemod := (exp, P) -> `if`(P=0, exp, exp mod P);
   B := 1;
   C := 1;
   L := 0;
   k := 1;
   b := 1;
   for n from 0 to 2*N-1 do
     d := s[n+1];
     for i from 1 to L do
       d := safemod(d + coeff(C,x^i)*s[n-i+1], P);
     od;
     if d=0 then k := k+1 fi;
     if (d <> 0 and 2*L > n) then
       C := safemod(expand(C - d*x^k*B/b), P);
       k := k+1;
     fi;
     if (d <> 0 and 2*L <= n) then
       T := C;
       C := safemod(expand(C - d*x^k*B/b), P);
       B := T;
       L := n+1-L;
       k := 1;
       b := d;
     fi;
   od;
   return C;
end:
```

Here is an example: let's take the following simple bitstream $\{u_n\}$ of length 30:

$$Z := [0,0,0,1,0,0,1,1,0,1,0,1,1,1,1,1,0,0,0,1,0,0,1,1,0,1,0,1,1,1,1].$$

Then the execution of the Berlekamp–Massey procedure **BM(Z,N,2,x)**; for $N = 4, 5, \dots, 15$ produces the polynomial

$$x^4 + x^3 + 1.$$

To understand what it stands for, imagine it as a code for the linear recurrence satisfied by $Z$. To see this recurrence, solve for 1 in the above polynomial, imagining that it equals zero in $\mathbb{Z}/2\mathbb{Z}$:

$$1 = x^3 + x^4.$$

Now imagine the left-hand side of the above equality as standing for a term $u_n$, while the monomials on the right-hand side, looked at in a "negative lens," stand for $u_{n-3}$ and $u_{n-4}$. Thus the linear recurrence satisfied by $Z$ will be

$$u_n = u_{n-3} + u_{n-4}$$

starting from the initial conditions (seed)

$$0, 0, 0, 1.$$

So, we can say that the linear complexity of the binary sequence $Z$ is 4 (and it will remain 4 if $Z$ is extended indefinitely to an infinite binary sequence of period 15).

On the other hand, if $N = 3$, the **BM(Z,N,2,x)**; produces $x^4 + 1$, indicating that the recurrence relation $u_n = u_{n-4}$ is the right minimal output for the initial segment $[0, 0, 0, 1, 0, 0]$ of $Z$ of length $2N = 6$. Finally, if $N = 16$, then **BM(Z,N,2,x)**; gives an error (recall that in order for **BM(s,N,2,x)**; to work, the length of $s$ must be at least $2N$).

Bitstreams naturally associated to the Legendre character of primes behave very well from the point of view of linear complexity, as was shown in the 1998 paper (Ding et al. 1998). The sequences $s_p^\infty$ considered there are infinite, consisting of repeated binary blocks of length $p$, and defined as follows:

$$u_n = \begin{cases} \left(1 + \left(\frac{n}{p}\right)\right)/2, & \text{if } n \neq 0 \bmod p \\ 0, & \text{if } n \neq 0 \bmod p \end{cases}.$$

The complete characterization of the linear complexities $L\left(s_p^\infty\right)$ of such 0/1 sequences provided in Ding et al. (1998) is given below:

$$L\left(s_p^\infty\right) = \begin{cases} (p+1)/2, & \text{if } p = 8t - 1 \\ (p-1)/2, & \text{if } p = 8t + 1 \\ p, & \text{if } p = 8t + 3 \\ p - 1, & \text{if } p = 8t + 3 \end{cases}$$

To take an example, let's say that $p = 223$. This is a prime of the form $8t - 1$, so the linear complexity $L(s_{223}^\infty)$ obtained by repeating the canonical bit string associated to the quadratic character modulo 223 should be $(223 + 1)/2 = 112$. Let's determine this through the MAPLE Berlekamp–Massey procedure BM shown above.

But how can we deal with such infinite bitstreams? It's simple: just consider repeating twice the period of $s_{223}^{\infty}$,

```
> p:=223; L1:=[seq((1+legendre(k,p))/2,k=1..p-1)]:
L2:=[0,op(L1),0,op(L1)];
```

and apply BM to the "doubled list" L2 of length $2p$:

```
> BM(L2,p,2,x);
```

This will give us a feedback polynomial of degree 112 of the linear feedback shift register (LFSR) generating $s_{223}^{\infty}$, its degree being the linear complexity as determined formally in Ding et al. (1998):

$$x^{112} + x^{109} + x^{107} + x^{106} + x^{101} + x^{100} + x^{97} + x^{94} + x^{93} + x^{92} + x^{86} + x^{85} + x^{84} + x^{81}$$
$$+ x^{80} + x^{77} + x^{73} + x^{72} + x^{67} + x^{66} + x^{65} + x^{64} + x^{63} + x^{62} + x^{61} + x^{60} + x^{59}$$
$$+ x^{56} + x^{52} + x^{50} + x^{46} + x^{44} + x^{43} + x^{40} + x^{39} + x^{38} + x^{37} + x^{35} + x^{33} + x^{31}$$
$$+ x^{30} + x^{29} + x^{28} + x^{27} + x^{26} + x^{25} + x^{23} + x^{21} + x^{20} + x^{16} + x^{14} + x^{13} + x^{11} + x^{9}$$
$$+ x^{6} + x^{5} + x^{4} + x^{3} + x + 1$$

Let us return to bitstreams obtained from GPF sequences and see whether we can replicate such good linear complexities. For the linear complexity of the finite 0/1 sequence $s$ corresponding to the above greatest prime factor sequence (1) of length 101, the Berlekamp–Massey procedure for $N = 50$, **BM(s,50,2,x)**; produces a feedback polynomial of degree 55:

$$x^{55} + x^{53} + x^{50} + x^{49} + x^{47} + x^{46} + x^{45} + x^{44} + x^{43} + x^{41} + x^{40}$$
$$+ x^{37} + x^{36} + x^{34} + x^{31} + x^{29} + x^{28} + x^{24} + x^{21} + x^{20} + x^{18}$$
$$+ x^{16} + x^{15} + x^{13} + x^{12} + x^{11} + x^{9} + x^{8} + x^{7} + x^{4} + x^{3} + x^{2} + x^{1}$$

As for the linear complexity of the infinite sequence obtained by repeating $s$ again and again, it will be enough, as discussed previously, to use the BM procedure for a duplicate of $s$ into **s2: = [op(s),op(s)]**;. We thus get **BM(s2,101,2,x)**;, indicating a feedback polynomial of degree 100 (hence a linear complexity of 100).

Let's now take a finite 0/1 sequence that corresponds to a randomly chosen 1000-term GPF sequence, say of order 3. In most cases we find a good linear complexity close to 500.

```
> a:=rand(1..K)(); b:=rand(1..K)(); c:=rand(1..K)();
f:=(x1,x2,x3)->(x2,x3,P(c*x1+b*x2+a*x3)):
x(0):=ithprime(rand(1..W)()); x(1):=ithprime(rand(1..W)());
x(2):=ithprime(rand(1..W)()); for r from 3 to 999 do
x(r):=P(a*x(r-1)+b*x(r-2)+c*x(r-3)) end do:
LP:=[seq(gpf2stream(x(r)),r=0..999)]: degree(BM(LP,500,2,x));
```

A typical run of the above code may pick some GPF recurrence coefficients and initial conditions of the form

$$\begin{cases} (a,b,c) = (9,7,8) \\ (x_0,x_1,x_2) = (7703,2027,3469) \end{cases}$$

This will be followed by the Berlekamp–Massey procedure for the associated 0/1 sequence $\{u_n\}_{n=0}^{999}$, which produces a polynomial of degree 499. This means that in most cases, even for bitstreams of a limited length 1000 based on GPF sequences of order 3, the corresponding bit string has a linear complexity close to $1000/2 = 500$.

We performed 100 such random trials and visualized the data for

$$\text{"LINEAR COMPLEXITY}/1000\text{"}$$

in histogram form.

```
> TRIALS:=1000: for T from 1 to TRIALS do W:=1000: K:=10:
a:=rand(1..K)(): b:=rand(1..K)(): c:=rand(1..K)():
f:=(x1,x2,x3)->(x2,x3,P(c*x1+b*x2+a*x3)):
x(0):=ithprime(rand(1..W)()): x(1):=ithprime(rand(1..W)()):
x(2):=ithprime(rand(1..W)()): for r from 3 to 999 do
x(r):=P(a*x(r-1)+b*x(r-2)+c*x(r-3)) end do:
LP:=[seq(gpf2stream(x(r)),r=0..999)]:
LC(T):=evalf(degree(BM(LP,500,2,x))/1000): end do:
BMSEQ:=[seq(LC(T),T=1..TRIALS)]: Histogram(BMSEQ);
```

The result was the following:

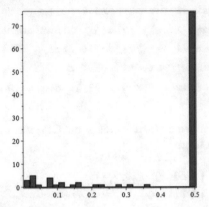

This means that most linear complexities for such 1000-term 0/1 strings are close to 500, which is encouraging. Another set of 1000 random trials (based on the same type of GPF sequences, of third order and length 1000) produces a similar histogram (more than 70% of the generating strings have linear complexity very close to half of their length):

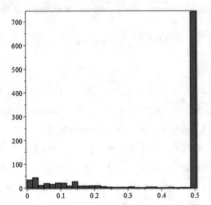

Now, if in our randomized search we increase the maximum size of the coefficients $a, b, c$ in the GPF recurrence and repeat the random experiment, it gets better:

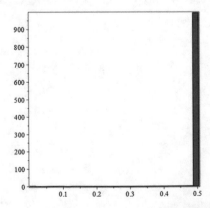

In more detail, the MAPLE **TallyInto** option

```
> TallyInto(BMSEQ,default,bins=50);
```

shows that in 989 out of 1000 trials, the linear complexity of the bitstreams produced from GPF sequences divided by their length (1000) lies in the interval $[0.49394, 0.50400]$.

We expect the linear complexity of such bitstreams to improve even more once we get to work with GPF sequences of higher order that have larger periods and/or preperiods.

Assuming the ultimate periodicity conjecture for GPF sequences $\{x_n\}_{n \geq 0}$, we will perform an analysis of the bitstreams generated by transforming their periods

$$[x_{\text{PREPERIOD}}, x_{\text{PREPERIOD}+1}, \ldots, x_{\text{PREPERIOD}+\text{PERIOD}-1}]$$

into 0/1 sequences as previously shown,

$$U := [u_{\text{PREPERIOD}}, u_{\text{PREPERIOD}+1}, \ldots, u_{\text{PREPERIOD}+\text{PERIOD}-1}],$$

where $u_j := \text{gpf2stream}(x_j)$, and then repeating the terms of above periodic unit (binary block) in order to obtain an infinite 0/1 sequence $U$:

$$U^\infty(\{x_n\}) := [u_{\text{PREPERIOD}}, u_{\text{PREPERIOD}+1}, \ldots, u_{\text{PREPERIOD}+\text{PERIOD}-1},$$
$$u_{\text{PREPERIOD}}, u_{\text{PREPERIOD}+1}, \ldots]$$

In other words, to every GPF sequence $\{x_n\}_{n \geq 0}$ we associate the above periodic bitstream, in the hope that we will have thus produced a reasonably well behaved stream cipher, with a good linear complexity. As noticed before, to find the linear complexity $L(U^\infty(\{x_n\}))$ of the stream $U^\infty(\{x_n\})$, it will be enough to apply the Berlekamp–Massey algorithm to the 0/1 sequence $[U, U]$ of length $2 \cdot \text{PERIOD}$ obtained by duplicating $U$, followed by the application of the procedure BM:

$$L(U^\infty(\{x_n\})) = \text{BM}([U, U], \text{PERIOD}, 2, x).$$

The hope translates into having $L(U^\infty(\{x_n\}))$ as close to the quantity PERIOD as possible. To validate that hope, the plan is (at least as a first attempt) to randomly pick a GPF sequence $\{x_n\}_{n \geq 0}$ of a fairly long period and calculate $L(U^\infty(\{x_n\}))$ by applying the BM procedure.

Our "first attempt" was the randomly chosen GPF sequence $\{x_n\}_{n \geq 0}$ defined as follows:

$$\begin{cases} x_n = P(5x_{n-1} + 5x_{n-2} + 3x_{n-x} + 5x_{n-4}) \\ x_0 = 67, x_1 = 53, x_2 = 97, x_3 = 67 \end{cases}$$

For this particular GPF sequence, Floyd's cycle-finding algorithm produces a preperiod of 5843 and a period of 1213. The bit string $U$ shown below for the record is produced from a GPF bitstream $U^{\infty}(\{x_n\})$ that has excellent linear complexity, namely $L(U^{\infty}(\{x_n\})) = 1213 = \text{PERIOD}$, exactly half the length of the duplicate $[U, U]$ with feedback polynomial $x^{1213} + 1$. This means that if we look for a summary LFSR presentation of the bitstream $U^{\infty}(\{x_n\})$, we can't do any better than listing the full period block $U$ (of length 1213) and then stating that $U$ repeats again and again:

$U = [0, 0, 1, 0, 1, 1, 0, 1, 0, 0, 0, 0, 1, 1, 0, 0, 1, 0, 1, 1, 1, 1, 0, 0, 1, 0, 0, 0, 1, 0, 1, 1, 1,$
$\quad 1, 1, 0, 0, 1, 1, 1, 0, 1, 1, 1, 0, 1, 1, 0, 1, 1, 1, 0, 1, 1, 0, 1, 1, 0, 1, 1, 0, 0, 1, 0, 0, 1, 0,$
$\quad 1, 0, 1, 1, 1, 1, 0, 1, 1, 0, 0, 0, 0, 0, 1, 0, 1, 0, 1, 0, 1, 1, 0, 0, 0, 0, 1, 0, 1, 1, 0, 1, 1, 0,$
$\quad 1, 0, 0, 1, 0, 1, 0, 1, 1, 0, 1, 0, 0, 1, 1, 0, 0, 0, 0, 0, 1, 0, 1, 1, 1, 0, 1, 0, 1, 0, 0, 1, 1, 1,$
$\quad 1, 1, 1, 1, 0, 1, 0, 1, 0, 0, 0, 0, 1, 0, 0, 0, 1, 0, 1, 1, 1, 0, 1, 0, 0, 0, 1, 0, 0, 0, 1, 1,$
$\quad 1, 0, 1, 0, 0, 1, 1, 1, 0, 0, 0, 1, 1, 0, 1, 1, 1, 0, 0, 0, 1, 1, 1, 0, 1, 1, 1, 0, 1, 1, 0, 0, 0, 0, 1,$
$\quad 0, 1, 0, 0, 0, 1, 1, 1, 0, 0, 0, 0, 1, 0, 1, 0, 1, 1, 0, 1, 1, 1, 0, 0, 1, 0, 0, 0, 1, 1, 1, 0, 0, 0,$
$\quad 0, 1, 1, 0, 0, 1, 0, 0, 1, 1, 0, 1, 1, 0, 0, 0, 1, 1, 0, 1, 1, 1, 0, 1, 1, 0, 1, 0, 1, 0, 1, 1, 0, 0,$
$\quad 1, 0, 1, 0, 1, 0, 1, 0, 0, 0, 0, 0, 1, 0, 0, 0, 1, 0, 1, 0, 1, 0, 1, 1, 1, 1, 1, 1, 0, 1, 1, 1, 1, 1,$
$\quad 0, 0, 1, 0, 1, 1, 1, 0, 1, 0, 0, 0, 0, 0, 1, 0, 1, 0, 0, 0, 1, 1, 1, 1, 1, 1, 1, 1, 1, 0, 1, 1, 0, 1, 1, 1,$
$\quad 1, 0, 0, 0, 1, 1, 1, 0, 0, 0, 0, 1, 1, 1, 1, 1, 0, 0, 0, 1, 1, 0, 0, 1, 1, 1, 1, 1, 0, 0, 0, 0, 1, 1, 1,$
$\quad 1, 1, 1, 1, 0, 0, 1, 1, 1, 1, 1, 1, 1, 1, 0, 1, 1, 1, 1, 1, 1, 0, 0, 1, 0, 0, 1, 1, 0, 0, 0, 0, 1, 1,$
$\quad 1, 0, 1, 1, 1, 0, 1, 0, 1, 1, 1, 0, 0, 1, 0, 1, 1, 1, 1, 0, 0, 0, 0, 0, 1, 0, 0, 1, 1, 1, 1, 1, 0, 1, 1,$
$\quad 0, 0, 1, 1, 0, 0, 0, 1, 0, 1, 0, 1, 0, 1, 0, 1, 1, 0, 1, 1, 1, 1, 1, 0, 0, 0, 0, 1, 0, 1, 0, 1, 1, 1,$
$\quad 1, 1, 0, 1, 1, 0, 1, 0, 0, 0, 0, 1, 0, 0, 1, 1, 0, 1, 0, 0, 1, 1, 1, 1, 1, 0, 1, 0, 0, 1, 1, 1, 0, 0,$
$\quad 1, 0, 1, 0, 1, 0, 0, 0, 1, 1, 1, 1, 0, 0, 0, 0, 0, 0, 1, 0, 0, 0, 0, 1, 1, 0, 1, 1, 0, 0, 0, 0, 0, 1,$
$\quad 0, 1, 1, 1, 0, 1, 0, 0, 1, 1, 1, 0, 0, 0, 1, 0, 0, 1, 0, 0, 1, 0, 1, 1, 0, 1, 1, 1, 1, 1, 1, 0, 0, 1,$
$\quad 0, 1, 0, 0, 1, 1, 1, 0, 0, 0, 0, 1, 0, 0, 0, 0, 0, 1, 0, 1, 1, 0, 1, 1, 0, 1, 1, 0, 0, 1, 0, 1, 0,$
$\quad 1, 0, 0, 1, 0, 1, 0, 1, 0, 1, 0, 0, 0, 1, 1, 1, 1, 1, 1, 0, 0, 0, 0, 0, 0, 0, 1, 1, 0, 1, 1, 0, 1, 1,$
$\quad 0, 1, 1, 0, 1, 1, 1, 0, 0, 0, 0, 1, 0, 1, 1, 0, 1, 0, 1, 0, 0, 1, 0, 1, 1, 0, 1, 0, 0, 1, 1, 1, 0,$
$\quad 1, 1, 1, 1, 1, 1, 0, 0, 0, 1, 1, 1, 1, 1, 1, 1, 0, 1, 0, 0, 1, 0, 1, 1, 0, 0, 0, 0, 0, 0, 1, 1, 1, 1,$
$\quad 0, 1, 0, 0, 0, 0, 0, 1, 0, 1, 0, 0, 0, 1, 0, 0, 1, 1, 1, 1, 0, 0, 1, 0, 0, 1, 0, 0, 1, 0, 1, 1, 1, 0,$
$\quad 0, 1, 1, 0, 1, 1, 1, 1, 1, 1, 0, 1, 1, 1, 0, 0, 1, 1, 0, 0, 1, 1, 1, 0, 1, 0, 0, 1, 1, 1, 0, 1, 0, 0,$
$\quad 0, 0, 0, 1, 1, 1, 1, 1, 0, 0, 1, 1, 1, 0, 0, 1, 1, 0, 1, 0, 0, 1, 0, 0, 0, 0, 1, 1, 1, 1, 1, 0, 1, 0,$
$\quad 1, 1, 1, 0, 0, 0, 1, 1, 1, 1, 0, 1, 0, 1, 0, 0, 0, 0, 0, 1, 0, 0, 0, 1, 1, 1, 1, 1, 1, 1, 1, 1, 1, 1, 1,$
$\quad 0, 1, 1, 1, 0, 0, 1, 0, 1, 0, 1, 0, 0, 1, 0, 0, 1, 0, 1, 0, 1, 1, 1, 0, 0, 1, 1, 1, 1, 0, 0, 1, 0,$
$\quad 1, 1, 0, 1, 1, 0, 0, 1, 0, 1, 1, 1, 0, 0, 1, 0, 0, 1, 0, 1, 1, 0, 0, 0, 0, 1, 0, 0, 1, 0, 0, 0, 1,$
$\quad 1, 0, 0, 1, 1, 1, 0, 0, 1, 1, 1, 1, 0, 1, 1, 0, 0, 1, 0, 1, 0, 1, 1, 0, 1, 1, 0, 0, 0, 0, 1, 1, 0,$
$\quad 0, 0, 0, 1, 0, 1, 1, 1, 1, 0, 0, 0, 1, 0, 0, 0, 0, 1, 1, 1, 0, 1, 0, 0, 1, 0, 0, 0, 0, 1, 0, 0,$

0, 0, 1, 1, 1, 0, 1, 0, 1, 0, 0, 1, 1, 0, 0, 0, 1, 1, 1, 1, 0, 0, 1, 0, 0, 1, 1, 1, 1, 1, 0, 1,
0, 0, 0, 0, 0, 1, 0, 0, 0, 1, 1, 0, 0, 1, 1, 1, 0, 0, 0, 1, 0, 1, 0, 1, 1, 1, 1, 0, 0, 0, 0, 0,
1, 0, 0, 1, 0, 0, 1, 0, 1, 0, 0, 1, 0, 1, 1, 0, 1, 0, 1, 0, 1, 0, 0, 0, 0, 0, 0, 1, 1, 0, 1, 1, 1, 1, 1,
1, 1, 1, 0, 1, 0, 0, 0, 1, 1, 0, 0, 0, 1, 0, 1, 0, 0, 0, 0, 1, 0, 0, 1, 1, 1, 1, 0, 1, 0, 0, 1, 1,
1, 0, 0, 1, 0, 0, 1, 0, 0, 1, 1, 0, 1, 1, 0, 1, 1, 0, 1, 0, 1, 1, 0, 0, 0, 1, 1, 0, 0, 0, 0, 0, 1,
0, 0, 1, 0, 0, 0, 1, 0, 0, 0, 0, 0, 0, 1, 0, 0, 0, 1, 1, 1, 1, 1, 0, 1, 0, 0, 0, 0, 1, 0, 0, 1,
1, 1, 1, 1, 0, 1, 0, 0, 0, 0, 0, 0, 1, 1, 0, 0, 0, 0, 0, 1, 1, 1, 0, 1, 0, 1, 1, 1, 0, 1, 1, 0, 1, 1, 1, 1, 1]

Interestingly enough, in the same particular case, the Berlekamp–Massey algorithm applied to the "preperiod-related" 0/1 sequence of length 5843 obtained from the terms of the same GPF sequence $\{x_n\}_{n \geq 0}$ that are not in the limit cycle, that is,

$$\text{PRE} := [u_0, u_1, \ldots, u_{\text{PREPERIOD}-1}] = [u_0, u_1, \ldots, u_{5842}],$$

again yields an excellent linear complexity close to the half of the preperiod, with the polynomial

$$\text{BM(PRE}, 2921, 1, x)$$

having degree 2920.

Note that if we apply the Berlekamp–Massey algorithm for the finite bit string $U$ (of length 1213), we will get 607.

An excellent, freely available, Python-based online tool for calculating the linear complexity of a finite binary string using the Berlekamp–Massey algorithm (powered by the Google App Engine) is available at http://berlekamp-massey-algorithm.appspot.com/.

There is, though, a difference between the above online tool and the MAPLE procedure that we applied here, in that the polynomial results are reciprocal to each other. Note that if you want to use the online tool to calculate the linear complexity of an infinite periodic bitstream obtained by repeating the block $U$, just enter a duplicate of $U$.

We were (and are) especially interested in linear complexities of infinite periodic bitstreams obtained by repeating a block of bits $U$ appearing in the 0/1 sequence associated to a greatest prime factor sequence $\{x_n\}_{n \geq 0}$, as shown above.

Usually we worked with $U^\infty(\{x_n\})$, where we take $U$ to be the block of bits corresponding to the limit cycle of $\{x_n\}_{n \geq 0}$.

For this purpose, we can use the following MAPLE procedure written for GPF sequences of order four (it can also be adapted to different orders):

```
BMGPF:=proc(a::integer, b::integer, c::integer, d::integer,
x0::integer, x1::integer, x2::integer, x3::integer);
f:=(x1,x2,x3,x4)->(x2,x3,x4,P(d*x1+c*x2+b*x3+a*x4)):
x(0):=x0; x(1):=x1; x(2):=x2; x(3):=x3;
SEED:=(x(0),x(1),x(2),x(3)):
X:=SEED: SLOW:=f(X): FAST:=f(f(X)): k:=1: while SLOW<>FAST do
SLOW:=f(SLOW): FAST:=f(f(FAST)): k:=k+1; end do:
TENTATIVE_START:=k:
Y:=SLOW: X:=SLOW: SLOW:=f(X): FAST:=f(f(X)): k:=1: while
SLOW<>FAST do SLOW:=f(SLOW): FAST:=f(f(FAST)): k:=k+1; end do:
PERIOD:=k;
m:=0: U:=SEED: V:=Y: while U<>V do m:=m+1; U:=f(U); V:=f(V) end
do: PREPERIOD:=m;
for r from 4 to PREPERIOD+2*PERIOD do x(r):=P(a*x(r-1)+b*x(r-
2)+c*x(r-3)+d*x(r-4)) end do:
LP:=[seq(gpf2stream(x(r)),r=PREPERIOD..PREPERIOD+2*PERIOD-1)]:
BMGPF:=(PREPERIOD,PERIOD,degree(BM(LP,PERIOD,2,x)));
end proc;
```

Note the fairly amazing fact (at least at the beginning of this experience) that the GPF sequences $\{x_n\}_{n\geq 0}$ on which the generated infinite bitstreams $U^\infty(\{x_n\})$ are based have small order (order four in the above procedure), which contrasts with the fact that the associated bitstreams have a large linear complexity (that is, we need linear recurrences of large order in order to describe them).

At the same time, $\{x_n\}_{n\geq 0}$ remains hidden from the observed bitstream.

Let's try some more examples.

- For **BMGPF(1,1,1,6,2,3,5,7)**; it doesn't take much time to obtain the result, in the form of the triple of numbers representing the preperiod, period, and the linear complexity of the infinite bitstream $U^\infty(\{x_n\})$ obtained by repeating the block of bits $U$ obtained by applying the modular procedure **gpf2stream** to the elements of the limit cycle $\{x_n\}_{n=25591}^{34613}$:

$$25591, 9023, 9022.$$

Therefore, the period of the GPF sequence

$$\begin{cases} x_n = P(x_{n-1} + x_{n-2} + x_{n-3} + 6x_{n-4}) \\ x_0 = 2, x_1 = 3, x_2 = 5, x_3 = 7 \end{cases}$$

is 9023, and the linear complexity of $U^\infty(\{x_n\})$ is 9022.
- Trying **BMGPF(2,1,1,1,11,37,2,73)**; yields a preperiod of 3277, a period of 5127, and a linear complexity of 5126 for the aforementioned stream $U^\infty(\{x_n\})$ based on the fourth-order GPF sequence

$$\begin{cases} x_n = P(2x_{n-1} + x_{n-2} + x_{n-3} + x_{n-4}) \\ x_0 = 11, x_1 = 37, x_2 = 2, x_3 = 73 \end{cases}$$

We can use MAPLE to generate the finite 0/1 block of length 5127 $U = [x_{3277}, \ldots, x_{8403}]$ that constitutes the period of $U^\infty(\{x_n\})$. By selecting the list $U$, among the options that MAPLE presents is *kernel density plot*, which provides a smooth probability density estimate of a random variable that could, hypothetically, produce $U$. In this case, we get the following distribution:

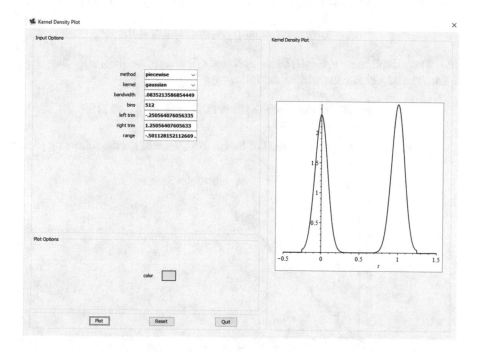

The fact that the bin around 1 is slightly higher than the bin around 0 illustrates again the curious "Chebyshev-like bias" that we noticed for the $\pm1$ walks associated to GPF sequences.

- Let us adapt BMGPF to third-order GPF sequences and use the resulting procedure for the following special GPF-tribonacci sequence (see Section 3.2):

$$\begin{cases} x_n = P(x_{n-1} + x_{n-2} + x_{n-3}) \\ x_0 = 2, x_1 = 3, x_2 = 5 \end{cases}.$$

We will get a preperiod of 1, a period of 100, while the linear complexity of the infinite stream cipher $U^\infty(\{x_n\})$ is 100. Here is the 0/1 period block of $U^\infty(\{x_n\})$:

$$U := [1,0,0,0,1,0,1,0,1,1,1,1,1,1,0,1,0,0,1,0,1,1,1,1,0,1,1,1,1,0,$$
$$0,1,1,1,1,0,1,1,1,1,0,1,1,1,0,0,0,1,1,1,1,1,0,1,0,0,0,0,1,1,1,0,$$
$$0,1,0,1,1,1,1,1,1,0,1,1,1,0,1,0,0,1,1,0,1,1,1,0,1,1,0,0,0,1,1,1,$$
$$0,0,1,0,0,0].$$

Using the Berlekamp–Massey procedure, we can now visualize the linear complexity profile "LCP" of the infinite bitstream $U^{\infty}(\{x_n\})$ based on the greatest prime factor sequence $\{x_n\}_{n \geq 0}$. Say $U^{\infty}(\{x_n\})$ has the block $U$ as period, i.e.,

$$U = [\text{gpf2seq}(x_0), \ldots, \text{gpf2seq}(x_{m-1})].$$

The duplicate $V := [U, U] = [\text{gpf2seq}(x_0), \ldots, \text{gpf2seq}(x_{2m-1})]$ has length $2m$. The MAPLE code will be the following:

```
> LCP:=[seq(degree(BM(V,N,2,x)),N=0..m)]:
> listplot(LCP);
```

Based on the above 0/1 string $U$ of length 100, we will get the following picture:

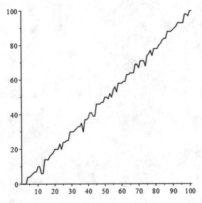

Note that if we would like to go further than $V = [U, U]$, the plot levels to a constant value equaling the linear complexity of the whole bitstream $U^{\infty}(\{x_n\})$. For example, if instead of $V$ we considered a $U$ in triplicate (that is, say $W = [U, U, U]$), the resulting plot of the values $BM(W, N2, x)$ for $0 \leq N \leq 150$ will be as shown below:

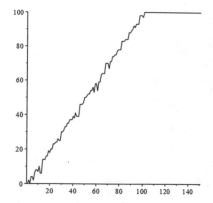

The linear complexity profile picture indicates that the initial segments of the bitstreams behave fairly well from a linear complexity point of view.

To summarize, the bitstreams $U^\infty(\{x_n\})$ generated, as indicated in this section, from greatest prime factor sequences $\{x_n\}_{n \geq 0}$ (even if they are defined by a recurrence of a relatively small order) appear to have large periods and large linear complexities, so that they potentially could make good candidates for stream ciphers used in one-time-pad cryptosystems.

In the calculation of the elements of the stream ciphers, the greatest prime sequence (including recurrence coefficients and seed) is kept private. We anticipate a computational problem, though, that may appear if we increase the order of the GPF recurrence (in which case, difficult factoring issues may arise).

All in all, we feel that this method of producing bitstreams has some potential in deriving good stream ciphers.

The topic of Chebyshev-like bias of $\pm 1$ walks associated to GPF sequences was part of a senior capstone project presented in May 2017 at Ohio Northern University (Haver 2017).

# Appendix A
# Review of Frequently Used Functions, Hands-On Visualization

## A.1 Headers

Here are the basic headers that we include (and load) at the head of the MAPLE worksheets that we used. Some were used frequently, others more rarely.

```
> with(numtheory): with(ListTools): with(plots):
with(StringTools): with(RandomTools): with(Statistics):
```

## A.2 Frequently Used Functions

And here are our main heroes, which we incorporate at the beginning of the worksheet whenever needed.

- The greatest prime factor (GPF) function

```
P:=n->max(1,op(factorset(n)));
```

- Conway's subprime function

```
C:=proc(n::integer) local u: if isprime(n)='true' or n=1 then
n else u:=factorset(n): n/min(seq(u[j], j=1..nops(u))) end if
end:
```

© Springer International Publishing AG 2017
M. Caragiu, *Sequential Experiments with Primes*,
DOI 10.1007/978-3-319-56762-4

## A.3    Visualizing Recurrent Sequences: Hands On, Brute Force

This is the first method used by us, and we keep using it in some cases, especially when we want to verify that we didn't introduce errors into a MAPLE code we wrote, making sure that it produces the correct period or preperiod. Let us take an example: consider the GPF sequence defined as follows:

$$\begin{cases} x_n := P(2x_{n-1} + 11x_{n-2} + 8) \\ x_n = 3, x_1 = 3 \end{cases}.$$

The GPF conjecture asserts that this prime sequence eventually enters a limit cycle. Let us see this cycle right away via direct MAPLE brute-force calculation and plotting:

```
x(0):=3; x(1):=13;
for r from 2 to 1000 do x(r):= P(2*x(r-1)+11*x(r-2)+8) end do:
X:=[seq(x(r),r=0..1000)]:
Listplot(X);
```

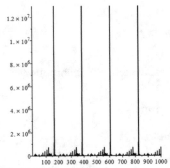

The periodicity is obvious from the plot.

Sometimes things need to be smoothed out, which translates into using a log-arithmic plot:

```
L:=[seq(ln(x(r)),r=0..1000)]:
Listplot(L);
```

This direct view is one of the "perks" of the activity of researching the greatest prime factor sequences or Conway subprime sequences. It is invaluable and direct, a "picture worth a thousand words."

And then there is another such perk: the "poor man's" way of getting the period from such a plot. It's a childish game of discovery that in some sense connects us

with the problem. Looking at the plot lets us see some terms of the sequence just
before the 400 mark and right after the 600 mark. Say we choose the following:

```
seq(x(r),r=385..395);
12982027, 2789, 147073, 2069, 419, 4721, 827, 53593, 449, 84347, 37
seq(x(r),r=600..610);
47087, 8831, 535627, 1168403, 64793, 12982027, 2789, 147073, 2069, 419, 4721
```

From the above listings, the sequence terms with subscripts 389 and 390 are
identical to the sequence terms with subscripts 609 and 610. Note that we need pairs
of consecutive terms to match, because we are dealing with a second-order
recurrence. The period is $609 - 389 = 220$. What is the preperiod?

## A.4   An Experiment with Angles

On a related note, we can talk about correlations between two vectors of the same
size, where periodic boundary conditions are assumed. Say the two vectors' pre-
sentations in list format are $L1 = [x_1, x_2, \ldots x_n]$ and $L2 = [y_1, y_2, \ldots y_n]$. Then the
cross-correlation coefficients are given by $\sum_{r=1}^{n} x_r y_{r+d}$, where $0 \leq d \leq n - 1$. Let
the periodic boundary conditions $\sum_{r=1}^{n} x_r y_{r+d}$, represent the inner product between
$L1$ and a $d$-step rotation of $L2$, with the corresponding angle $\theta_d$ given by

$$\cos(\theta_d) = \frac{\sum_{r=1}^{n} x_r y_{r+d}}{\sqrt{\sum_{r=1}^{n} x_r^2} \sqrt{\sum_{r=1}^{n} y_r^2}}.$$

We intend to place this in the context of our $\pm 1$ walks associated to greatest
prime factor sequences. If $L1$ and $L2$ are $\pm 1$ sequences, then the above cosine
simplifies to

$$\frac{1}{n} \sum_{r=1}^{n} x_r y_{r+d} \in [-1, 1].$$

We will work with two independent lists of GPF sequences of order 3, that is,
prime sequences $\{z_n\}_{n \geq 1}$ and $\{w_n\}_{n \geq 1}$ satisfying the recurrences

$$z_n = P(a_1 z_{n-1} + b_1 z_{n-2} + c_1 z_{n-3}) \text{ for } n \geq 4,$$
$$w_n = P(a_2 w_{n-1} + b_2 w_{n-2} + c_2 w_{n-3}) \text{ for } n \geq 4.$$

We will generate $V$ such sequences of each type. The recurrence coefficients $a_1, b_1, c_1, a_2, b_2, c_2$ are odd and otherwise randomly selected in a certain box $1 \leq a_1, b_1, c_1, a_2, b_2, c_2 \leq 2m - 1$, while the initial conditions $x_1, x_1, x_1, y_2, y_2, y_2$ are odd but otherwise randomly selected from among the primes $p_2, p_3, \ldots, p_w$.

For each such selection, the initial segments $\{z_n\}_{n=1}^{K}$ and $\{w_n\}_{n=1}^{K}$ of the generated GPF sequences $\{z_n\}_{n \geq 1}$ and $\{w_n\}_{n \geq 1}$ (respectively) will be converted to two $\pm 1$ vectors of size $K$ using the customary $z_n \mapsto 2 - (z_n \bmod 4)$ and $w_n \mapsto 2 - (z_n \bmod 4)$, respectively. Thus we obtain $V$ pairs of vectors with $\pm 1$ components and norm $\sqrt{K}$ each. Among each such pairs of vectors there are $K$ possible correlation coefficients, depending on the shift $d$.

In what follows we will work with a zero shift. Arguably a good (quasi)randomness indicator is a bell-shaped histogram centered at zero for the quantities (cosines)

$$\left( \sum_{r=1}^{K} z_r w_r \right) / K \in [-1, 1].$$

The experiment will be performed using the following MAPLE code, where we selected the parameters $V = 2000, K = 1000, M = 10, W = 100$.

```
V:=2000: K:=1000: M:=10: W:=100: for T from 1 to V do
a1:=2*rand(1..M)()-1: b1:=2*rand(1..M)()-1: c1:=2*rand(1..M)()-
1: x(1):=ithprime(rand(2..W)()): x(2):=ithprime(rand(2..W)()):
x(3):=ithprime(rand(2..W)()): for r from 4 to K do
x(r):=P(a1*x(r-1)+b1*x(r-2)+c1*x(r-3)) end do:
L1:=<seq(x(r),r=1..K)>: a2:=2*rand(1..M)()-1:
b2:=2*rand(1..M)()-1: c2:=2*rand(1..M)()-1:
x(1):=ithprime(rand(2..W)()): x(2):=ithprime(rand(2..W)()):
x(3):=ithprime(rand(2..W)()): for r from 4 to K do
x(r):=P(a2*x(r-1)+b2*x(r-2)+c2*x(r-3)) end do:
L2:=<seq(x(r),r=1..K)>: DP:=add(x,x in [seq((2-(L1[k] mod
4))*(2-(L2[k] mod 4)),k=1..K)]): A(T):=evalf(DP/K): end do:
COSDIST:=Histogram([seq(A(T),T=1..V)]);
```

The results detail an approximation of the distribution of angles between two initial segments resulting from third-order GPF sequences with random initial conditions and recurrence coefficients. The expected angle value is $\pi/2$, and the distribution looks indeed fairly narrow. The two histograms below are in two alternative formats (the MAPLE default and discrete with absolute frequency scale, respectively).

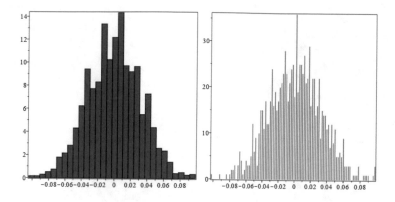

# Appendix B
# Review of Floyd's Algorithm
# and Floyd–Monte Carlo Data
# Acquiring for Periods

Below we include a MAPLE version of Floyd's cycle-finding algorithm (see Section 2.4 for details) for greatest prime factor sequences of order 4. We called this PERIOD4. It receives the recurrence constants and the seed components as input, and outputs the preperiod, period, and limit cycle.

```
PERIOD4 := proc(a::integer, b::integer, c::integer, d::integer, x0::integer, x1::integer, x2::
    integer, x3::integer)
    local f, SEED, X, SLOW, FAST, k, TENTATIVE_START, Y, PERIOD, m, U, V,
    PREPERIOD, r, LP, PERIOD4;
    f := (s, t, u, v) → (t, u, v, P(d*s + c*t + b*u + a*v));
    x(0) := x0;
    x(1) := x1;
    x(2) := x2;
    x(3) := x3;
    SEED := x0, x1, x2, x3;
    X := SEED;
    SLOW := f(X);
    FAST := f(f(X));
    k := 1;
    while SLOW <> FAST do
        SLOW := f(SLOW); FAST := f(f(FAST)); k := k + 1
    end do;
    TENTATIVE_START := k;
    Y := SLOW;
    X := SLOW;
    SLOW := f(X);
    FAST := f(f(X));
    k := 1;
    while SLOW <> FAST do
        SLOW := f(SLOW); FAST := f(f(FAST)); k := k + 1
    end do;
    PERIOD := k;
    m := 0;
    U := SEED;
    V := Y;
    while U <> V do m := m + 1; U := f(U); V := f(V) end do;
    PREPERIOD := m;
    for r from 4 to PREPERIOD + PERIOD − 1 do
        x(r) := P(a*x(r − 1) + b*x(r − 2) + c*x(r − 3) + d*x(r − 4))
    end do;
    LP := [seq(x(r), r = PREPERIOD..PREPERIOD + PERIOD − 1)];
    PERIOD4 := PREPERIOD, PERIOD, LP
end proc
```

© Springer International Publishing AG 2017
M. Caragiu, *Sequential Experiments with Primes*,
DOI 10.1007/978-3-319-56762-4

For example, **PERIOD4(1,1,1,1,2,3,5,7);** provides the period, preperiod, and the limit cycle for the GPF sequence

$$\begin{cases} x_n = P(x_{n-1}+x_{n-2}+x_{n-3}+x_{n-4}), \\ x(0) = 2, x(1) = 3, x(2) = 5, x(3) = 7. \end{cases}$$

```
> PERIOD4(1,1,1,1,2,3,5,7);
```
$$240, 14, [7, 5, 37, 17, 11, 7, 3, 19, 5, 17, 11, 13, 23, 2]$$

For other cases, the reader can adapt the design of the MAPLE code accordingly.

The "Floyd–Monte Carlo" idea comes from the fact that a greatest prime factor sequence of a Conway subprime sequence with a fixed recurrence relation can enter various limit cycles, depending on the initial conditions (the "seed" chosen). This method is supposed to produce two kinds of output:

- A list with the possible limit cycles lengths $T_1, T_2, \ldots, T_p$ found as a result of randomly varying the initial conditions within a preassigned "box" of the form $A \leq x_0, x_1, \ldots, x_m \leq B$.
- A list of proportions $pr_1, pr_2, \ldots, pr_p$ of occurrences for each period discovered in the previous step (actually, the two steps progress concurrently): $pr_1 + pr_2 + \cdots + pr_p = 1$.

In this way, we can obtain an approximate "picture" of the space of possible periods for the recurrence considered. Among the limitations of the method is that some periods may be very rare and may not appear unless we repeat the trials or increase the initial condition box size (which may result, though, in a prohibitively large running time). Also, there may exist two limit cycles of the same size that are otherwise disjoint.

Other than that, it is an exciting method of period discovery, and it can be applied to integer sequences or integer vector sequences (as we did with the Conway subprime Ducci games in Section 4.4).

For example let's say we want to look into the period distribution of the Conway subprime recurrence

$$x_n = C(2x_{n-1}+x_{n-2}).$$

We will let the initial conditions $SEED := (x_0, x_1)$ for the above recurrence vary in a box

$$1 \leq x_0, x_1 \leq K.$$

A number, $V$, of trials will be executed. Every such trial starts with choosing a random seed satisfying the above boxlike constraint, followed by the execution of the "period" part of Floyd's algorithm in order to derive the corresponding period PERSTAT (T). Then:

- We organize the periods the format of a list $L$ with $V$ elements.
- We convert $L$ into a set $S$ for a concise picture of the periods discovered.
- For each element of the set S, we count its number of occurrences in the list $L$. this will provide numerical information about the distribution of the possible periods.

```
V:=10000: a:=2: b:=1: K:=10000: f:=(x1,x2)->(x2,C(b*x1+a*x2)): for T
from 1 to V do SEED:=(rand(1..K)(),rand(1..K)()): X:=SEED:
SLOW:=f(X): FAST:=f(f(X)): k:=1: while SLOW<>FAST do SLOW:=f(SLOW):
FAST:=f(f(FAST)): k:=k+1; end do: TENTATIVE_START:=k: Y:=SLOW:
X:=SLOW: SLOW:=f(X): FAST:=f(f(X)): k:=1: while SLOW<>FAST do
SLOW:=f(SLOW): FAST:=f(f(FAST)): k:=k+1; end do: PERIOD:=k:
PERSTAT(T):=PERIOD: end do: L:=[seq(PERSTAT(T),T=1..V)]:
S:=convert(L,set); u:=numelems(S): for k from 1 to u do
FREQ(k):=Occurrences(S[k],L); end do: OCC:=[seq(FREQ(k),k=1..u)];
PROP:=seq(evalf(FREQ(k)/V),k=1..u);
```

We executed the above "Floyd–Monte Carlo" MAPLE code a couple of times with the above parameters, obtaining the following:

$$S := \{1, 7, 8, 11, 119, 231\}$$
$$OCC := [15, 1, 3, 14, 48, 9919]$$
$$PROP := 0.001500000000, 0.0001000000000, 0.0003000000000, 0.001400000000, 0.004800000000, 0.9919000000$$

$$S := \{1, 7, 8, 11, 119, 231\}$$
$$OCC := [21, 1, 1, 12, 52, 9913]$$
$$PROP := 0.002100000000, 0.0001000000000, 0.0001000000000, 0.001200000000, 0.005200000000, 0.9913000000$$

Expanding the search box by setting K = 100,000 and increasing the number of searches to 20,000 does not produce anything substantially new. Here is a sample output, with a period of length 231 occurring in more than 99% of the random searches.

$$S := \{1, 7, 8, 11, 119, 231\}$$
$$OCC := [9, 5, 1, 22, 114, 19849]$$
$$PROP := 0.0004500000000, 0.0002500000000, 0.00005000000000, 0.001100000000, 0.005700000000, 0.9924500000$$

Below we include a Floyd–Monte Carlo formal procedure, this time for second-order greatest prime factor sequences, producing the raw data of cycle lengths discovered, and their respective numbers of occurrences.

```
MCGPF := proc(a::integer, b::integer, V::integer, K::integer)
    local f, T, SEED, X, SLOW, FAST, k, TENTATIVE_START, Y, PERIOD, L, S, u, OCC,
    PROP, MCGPF;
    f := (x1, x2) → (x2, P(b*x1 + a*x2));
    for T to V do
        SEED := rand(1 ..K)( ), rand(1 ..K)( );
        X := SEED;
        SLOW := f(X);
        FAST := f(f(X));
        k := 1;
        while SLOW <> FAST do
            SLOW := f(SLOW); FAST := f(f(FAST)); k := k + 1
        end do;
        TENTATIVE_START := k;
        Y := SLOW;
        X := SLOW;
        SLOW := f(X);
        FAST := f(f(X));
        k := 1;
        while SLOW <> FAST do
            SLOW := f(SLOW); FAST := f(f(FAST)); k := k + 1
        end do;
        PERIOD := k;
        PERSTAT(T) := PERIOD
    end do;
    L := [seq(PERSTAT(T), T = 1 ..V)];
    S := convert(L, set);
    u := numelems(S);
    for k to u do FREQ(k) := ListTools:-Occurrences(S[k], L) end do;
    OCC := [seq(FREQ(k), k = 1 ..u)];
    PROP := seq(evalf(FREQ(k) / V), k = 1 ..u);
    MCGPF := S, OCC
end proc
```

This does not mean that other possible periods do not exist—see the limitations mentioned before—or it may be that, hypothetically, some periods can be reached for "astronomically high" values of the components of the seed vector.

But of course this is not statistics: at a maximum, we are interested in a very precise and clear-cut mathematical object (the configuration of all possible cycles). This approach may be considered "low tech," although it is a reasonably fast path to unveiling part of the truth in the absence of a formal derivation. It is a very exciting method of discovery, which can partially make up for the limitations discussed.

# Appendix C
# Julia Programs Used in Exploring GPF and Conway Magmas

The Julia codes below were part of the Google computing exploration of the greatest prime factor magmas (Section 3.7) and Conway subprime magmas (Section 4.3) that resulted in the joint papers (Caragiu and Vicol 2016; Caragiu et al., to appear in Fibonacci Q.), respectively.

They were written and executed in a Google computing environment by Paul A. Vicol, M.Sc. student in computing science at Simon Fraser University at the time (he is presently a Ph.D. student in machine learning at the University of Toronto).

These computations were instrumental in finalizing the last steps in the calculations of the expanding sets $G_n$ for a selection of greatest prime factor magmas $(\Pi, f_{a,b})$ satisfying the necessary conditions for cyclicity, and for the calculations involving sets that are similarly generated in the Conway subprime magmas, thus strengthening the computational support to an interesting conjectural representation of the golden section in the context of Conway subprime magmas.

## C.1.  GPF Magmas

```
function primes(n::Int64)
  isprime = ones(Bool,n)
  isprime[1] = false
  for i in 2:int64(sqrt(n))
    if isprime[i]
      for j in (i*i):i:n
        isprime[j] = false
      end
    end
  end
  return filter(x -> isprime[x], 1:n)
end
```

© Springer International Publishing AG 2017
M. Caragiu, *Sequential Experiments with Primes*,
DOI 10.1007/978-3-319-56762-4

```julia
ps = primes(100000000)

function memoize(f)
  d = Dict{Int64, Any}()
  function memoized_f(n::Int64)
    if n in keys(d)
      #print("Cached!\n")
      return d[n]
    end

    #print("Computing...\n")
    res = f(n)
    d[n] = res
    return res
  end
  return memoized_f
end


function gpf3(n::Int64)
  factors = Int64[]
  prime_index = 1
  d = ps[prime_index]

  while n > 1
    while n % d == 0
      push!(factors, d)
      n /= d
    end
    prime_index += 1
    d = ps[prime_index]
    if d*d > n
      if n > 1
        push!(factors, n)
```

```julia
        break
      end
    end
  end
  return factors[end]
end

memo_gpf3 = memoize(gpf3)

function run_gpf(a, b, g, n)

  dirname = "results-a$a-b$b-g$g"
  mkdir(dirname)

  A = Set([g])

  for iteration in 1:n
    num_pairs = 0
    tic()
    new_A = Set(A)

    A_list = sort!([item for item in A])
    l = length(A_list)

    for i in 1:l
      for j in i:l
        num_pairs += 1
        p = A_list[i]
        q = A_list[j]
        result1 = memo_gpf3(a * p + b * q)
        result2 = memo_gpf3(a * q + b * p)
        push!(new_A, result1)
        push!(new_A, result2)
      end
```

```
        end

        sorted_A = sort!([item for item in new_A])

        consecutive_primes = 0
        for index in 1:length(sorted_A)
          if ps[index] != sorted_A[index]
            break
          end
          consecutive_primes += 1
        end

        println("ITERATION $iteration FINISHED")
        println("A_$iteration = $sorted_A")
        println("NUM ELEMENTS IN A_$iteration = $(length(new_A))")
        println("Consecutive primes = $consecutive_primes")
        println("num_pairs = $num_pairs")

        stats_file = open("$dirname/gpf-stats-output-a$a-b$b-g$g-i$iteration.txt", "w")
        elements_file = open("$dirname/gpf-elements-output-a$a-b$b-g$g-i$iteration.txt", "w")

        write(stats_file, "Iteration $iteration\n")
        write(stats_file, "Number of elements in A_$iteration: $(length(new_A))\n")
        write(stats_file, "Consecutive primes = $consecutive_primes")
        write(stats_file, "Number of pairs: $num_pairs\n")
        write(stats_file, "-------------------------------\n")

        write(elements_file, "Elements in iteration $iteration: $sorted_A\n")
        write(elements_file, "-------------------------------\n")

        close(stats_file)
        close(elements_file)

        A = new_A


    toc()
      end
    end

    run_gpf(27,3,2,6)
```

## C.2.    Conway Subprime Magmas

```
function primes(n::Int64)
  isprime = ones(Bool,n)
  isprime[1] = false
  for i in 2:int64(sqrt(n))
    if isprime[i]
      for j in (i*i):i:n
        isprime[j] = false
      end
    end
  end
  return filter(x -> isprime[x], 1:n)
end

ps = Set(primes(1000000000))


function run_lpf(n)
  dirname = "results"
  mkdir(dirname)

  old_set = Set([1])
  new_set = Set([2])
```

```julia
    constructed_set = Set{Int64}()

    lpf_dict = Dict{Int64,Int64}()

    for iteration in 1:n
        num_pairs = 0
        start_time = tic()

        for old_item in old_set
            for new_item in new_set
                num_pairs += 1

                s = old_item + new_item
                if s in ps
                    result = s
                else
                    if s % 2 == 0
                        lpf = 2
                    else
                        try
                            lpf = lpf_dict[s]
                        catch e
                            upper_search_bound = int64(sqrt(s))
                            for prime in ps
                                if s % prime == 0
                                    lpf = prime
                                    lpf_dict[s] = lpf
                                    break
                                elseif prime >= upper_search_bound
                                    lpf = s
                                    lpf_dict[s] = lpf
                                    break
                                end
                            end
```

```julia
          end
        end

        result = div(s,lpf)
      end

      if !(result in old_set)
        if !(result in new_set)
          push!(constructed_set, result)
        end
      end
    end
  end
end


new_list = sort!([item for item in new_set])
l = length(new_list)


for i in 1:(l-1)
  for j in i:l
    num_pairs += 1

    x = new_list[i]
    y = new_list[j]

    s = x + y
    if s in ps
      result = s
    else
      if s % 2 == 0
        lpf = 2
      else
        try
```

```julia
              lpf = lpf_dict[s]
          catch e
            upper_search_bound = int64(sqrt(s))
            for prime in ps
              if s % prime == 0
                lpf = prime
                lpf_dict[s] = lpf
                break
              elseif prime >= upper_search_bound
                lpf = s
                lpf_dict[s] = lpf
                break
              end
            end
          end

          result = div(s,lpf)
        end

        if !(result in old_set)
          if !(result in new_set)
            push!(constructed_set, result)
          end
        end
      end
    end

old_set = union(old_set, new_set)
new_set = constructed_set
constructed_set = Set{Int64}()

end_time = toc()
```

```julia
        all_list = sort!([item for item in union(old_set, new_set)])

        println("Iteration $iteration, num elems in old set = $(length(old_set))")
        println("Num constructed elements = $(length(new_set))")
        println("Num pairs = $(num_pairs)")
        println("------------------------------")

        stats_file = open("$dirname/lpf-stats-output-i$iteration.txt", "w")
        elements_file = open("$dirname/lpf-elements-output-i$iteration.txt", "w")

        write(stats_file, "Iteration $(iteration)\n")
        write(stats_file, "Number of elements in (old_set U new_set): $(length(old_set))\n")
        write(stats_file, "Number of newly constructed elements: $(length(new_set))\n")
        write(stats_file, "Total number of elements in iteration $iteration: $(length(old_set) +
length(new_set))\n")
        write(stats_file, "Number of pairs: $(num_pairs)\n")
        write(stats_file, "------------------------------\n")

        write(elements_file, "Elements in iteration $iteration: $all_list\n")
        write(elements_file, "------------------------------\n")

        close(stats_file)
        close(elements_file)
    end
end

run_lpf(35)
```

# Appendix D
# What's Next? Epilogue and Some Reflections

Now at the end, the question is, what's next?"

Quite a lot.

Is expecting a formal proof of the ultimate periodicity of GPF sequences or Conway subprime sequences as realistic an objective as expecting a proof of the $3x + 1$ problem?

And if the general ultimate periodicity conjecture is false, one cannot point to an alleged counterexample and say, this particular GPF (or Conway subprime) sequence is not periodic, since a supercomputer ran for weeks and couldn't get to a cycle.

Are there any heuristic arguments that suggest that prime sequences satisfying recurrences of the form $x_n = P(c_0 + c_1 x_{n-1} + c_2 x_{n-2} + \cdots + c_d x_{n-d})$ are ultimately periodic? We didn't find a satisfactory one, but here is a very raw attempt based on a recurrence order reduction coupled with a size estimate. Of course, much could go wrong, and it can get us only so far:

- If $c_0 + c_1 x_{n-1} + c_2 x_{n-2} + \cdots + c_d x_{n-d}$ is prime, then $x_n = c_0 + c_1 x_{n-1} + c_2 x_{n-2} + \cdots + c_d x_{n-d}$, in which case we chose the estimate $x_n$ as $x_n \approx A x_{n-1}$, where $A$ is a constant (although unspecified, we could think of $A$ as, for example, $A \approx \sum c_i$).
- If $c_0 + c_1 x_{n-1} + c_2 x_{n-2} + \cdots + c_d x_{n-d}$ is composite, then with the same assumptions as above, we will take $x_n$, the GPF of $c_0 + c_1 x_{n-1} + c_2 x_{n-2} + \cdots + c_d x_{n-d} \approx A x_{n-1}$, of the form $x_n \approx \sqrt{A x_{n-1}}$ with the same $A$ as before. Of course, the assumption is again bold and of a "one size fits all" type, but we will keep it anyway.
- So we have a recurrence degree reduction in place: the above raw heuristic got us down to the first-order sequence

$$
x_n = \begin{cases} A x_{n-1}, & \text{if } c_0 + c_1 x_{n-1} + c_2 x_{n-2} + \cdots + c_d x_{n-d} \approx A x_{n-1} \text{ is prime,} \\ \sqrt{A x_{n-1}}, & \text{if } c_0 + c_1 x_{n-1} + c_2 x_{n-2} + \cdots + c_d x_{n-d} \approx A x_{n-1} \text{ is composite.} \end{cases}
$$

© Springer International Publishing AG 2017
M. Caragiu, *Sequential Experiments with Primes*,
DOI 10.1007/978-3-319-56762-4

- Now, using the prime number theorem, we can rewrite the above recurrence as
  follows:

$$x_n = \begin{cases} Ax_{n-1}, & \text{with a probability of } \frac{1}{\ln Ax_{n-1}}, \\ \sqrt{Ax_{n-1}}, & \text{with a probability of } 1 - \frac{1}{\ln Ax_{n-1}}. \end{cases}$$

- If $\delta_n$ is the number of digits of $x_n$ and if $a$ is the number of digits of $A$, then the
  above recurrence can be modeled (again with generosity) as follows, where
  $K(\approx \ln 10)$ is a fixed constant:

$$\delta_n = \begin{cases} a + \delta_{n-1}, & \text{with probability } \frac{1}{K(a+\delta_{n-1})}, \\ \frac{a+\delta_{n-1}}{2}, & \text{with probability } 1 - \frac{1}{K(a+\delta_{n-1})}. \end{cases}$$

This could be viewed as a "hand-waving" probabilistic hint (which could be verified by simulation) as to *why* the periodicity holds: if $\delta_{n-1}$ is very large (as in "much larger than $a$"), then it is much more likely that we will proceed to the next term through the event of (*roughly*) *dividing by* 2, which causes a major decrease in size from $\delta_{n-1}$, while the alternative (adding the fairly negligible amount $a$) cannot be expected to happen for many times in a row due to the very small probability $1/(Ka + K\delta_{n-1})$ of the event "adding $a$." This will (probabilistically) soon force the term $\delta_n$ down from any sufficiently high upper bound, and as such, $\delta_n$ will be "probabilistically confined" to a finite initial interval of natural numbers, the confinement being with a higher probability, the larger the interval.

However, ultimately this hand-waving explanation cannot be accepted, since our sequences are first of all not real random processes, and most importantly, excessively strong assumptions have been made. It probably makes some sense after the fact. i.e., after we have seen that the ultimate periodicity of GPF sequences is verified in all computer experiments that have been carried out, but that is a circular argument.

Returning to real possible arguments involving GPF sequences, perhaps a fundamental result could be proved to the effect that if a limit cycle exists, then its elements are less than a certain theoretically derived bound expressed in terms of the degree of the recurrence, the coefficients of the recurrence relation, and the sizes of the components of the seed.

Another favorite conjecture of ours involves the connection between the process of generating the Conway subprime magma and the golden section: if proved, it would definitely enrich the spectrum of meaning of that important constant with a representation solely in terms of the set of prime numbers.

Let us finish with one last project, this time related to Conway's subprime function. To keep it on the fun side, Floyd's cycle-finding algorithm ran at the author's office computer for more than a week, before the user inadvertedly and suddenly logged off ☺.

**CEP21** Find the period of the "Conway subprime tetranacci sequence" defined as follows:

$$\begin{cases} x_n = C(x_{n-1} + x_{n-2} + x_{n-3} + x_{n-4}), \\ x_0 = x_1 = x_2 = 0, \ x_3 = 1. \end{cases}$$

# References

N.S. Aladov, On the distribution of quadratic residues and nonresidues of a prime number $p$ in the sequence1, 2, ... $p - 1$, Mat. Sb **18**, 61–75 (1896, in Russian)

D.G. Allen, Babylonian mathematics. Lectures on the History of Mathematics, Texas A&M University. http://www.math.tamu.edu/∼dallen/masters/egypt_babylon/babylon.pdf. Last accessed on 11 Oct, 2016

A.L. Furno, Cycles of differences of integers. J. Number Theory **13**, 255–261 (1981)

J. Alm, T. Herald, A note on prime Fibonacci sequences. Fibonacci Q. **54**(1), 55–58 (2016)

G. E. Andrews, *Number Theory*. Dover (1994)

T.M. Apostol, *Introduction to Analytic Number Theory. Undergraduate Texts in Mathematics* (Springer, Berlin, 1976)

E. Bach, J.O. Shallit, in *Algorithmic Number Theory*, vol. 1. Efficient Algorithms. (The MIT Press, 1996)

G. Back, M. Caragiu, The greatest prime factor and recurrent sequences. Fibonacci Q. **48**(4), 358–362 (2010)

A. Bager, Proposed problem E2883, Amer. Math. Mon. **87** (1980)

R.C. Baker, G. Harman, J. Pintz, The difference between consecutive primes. II. Proc. London Math. Soc. **83**(3), 532–562 (2001)

N. Baxter, M. Caragiu, Arithmetic properties if some special sums. JP J. Algebra, Number Theory Appl. **4**(3), 455–463 (2004)

F. Berto, J. Tagliabue, Cellular Automata, *The Stanford Encyclopedia of Philosophy* (Summer 2012 Edition), ed. by E.N. Zalta, http://plato.stanford.edu/archives/sum2012/entries/cellular-automata/

J. Bezanson, S. Karpinski, V.B. Shah, A. Edelman, Julia: a fast dynamical language for technical computing. arXiv:1209.5145v1 (2012)

P. Billingsley, Prime numbers and Brownian motion. Am. Math. Mon. **80**, 1099–1115 (1973)

L. Blum, M. Blum, M. Shub, A simple unpredictable pseudo-random number generator. SIAM J. Comput. **15**(2), 364–383 (1986)

A. Bogomolny, Patterns in Pascal's triangle. Interactive mathematics miscellany and puzzles. http://www.cut-the-knot.org/arithmetic/combinatorics/PascalTriangleProperties.shtml, Accessed 26 Oct 2016

P. Borwein, *Computational Excursions in Analysis and Number Theory (CMS Books in Mathematics)* (Springer, Berlin, 2002)

J. Brace, Traffic flow simulation with cellular automata. Senior Capstone Project, Ohio Northern University, May 2010 (advisor M. Caragiu)

D. Bressoud, S. Wagon, *A Course in Computational Number Theory (Textbooks in Mathematical Sciences)*, 1 edn. (Key College, 2000)

F. Breuer, E. Lötter, B. van der Merwe, Ducci-sequences and cyclotomic polynomials. Finite Fields Their Appl. **13**(2), 293–304 (2007)

© Springer International Publishing AG 2017
M. Caragiu, *Sequential Experiments with Primes*,
DOI 10.1007/978-3-319-56762-4

F. Breuer, Ducci sequences in higher dimensions. Electron. J. Comb Number Theory **7**, #A24 (2007)

G. Brockman, R.J. Zerr, Asymptotic behavior of certain Ducci sequences. Fibonacci Q. **45**(2), 155–163 (2007)

R. Brown, J.J. Merzel, Limiting behavior in Ducci sequences. Period.Math. Hung. **47**(1–2), 45–50 (2003)

V. Brun, La série 1/5+1/7+1/11+1/13+1/17+1/19+1/29+1/31+1/41+1/43+1/59+1/61+..., où les dénominateurs sont nombres premiers jumeaux est convergente ou finie. Bull. des Sci. Math. (in French) **43**, 100–104, 124–128 (1919)

J. Buchmann, *Introduction to cryptography (Undergraduate Texts in Mathematics)*, 2nd edn. (February 22, 2009)

J.P. Buhler, P. Stevenhagen (eds), *Algorithmic Number Theory: Lattices, Number Fields, Curves and Cryptography (Mathematical Sciences Research Institute Publications)*, 1st edn. (Cambridge University Press, Cambridge, 2008)

P. Burcsi, S. Czirbusz, G. Farkas, Computational investigation of Lehmer's totient problem. Ann. Univ. Sci. Budap. Rolando Eötvös, Sect. Comput. **35**, 43–49 (2011)

D.A. Burgess, The distribution of quadratic residues and non-residues. Mathematika **4**, 106–112 (1957)

C. Caldwell, The Prime Glossary: Gilbreath's conjecture, The Prime Pages, http://primes.utm.edu/glossary/page.php?sort=GilbreathsConjecture

N.J. Calkin, J.G. Stevens, D.M. Thomas, A characterization for the length of cycles of the N-number ducci game. Fibonacci Q. **43**(1), 53–59 (2005)

M. Caragiu, First-order non-definability for primitive roots. Rev. Roumaine Math. Pures Appl. **44** (1999), no. 2, 167–169 (2000)

M. Caragiu. Recurrences based on the greatest prime factor function. JP J. Algebra, Number Theory Appl. 19(2), 155-163 (2010)

M. Caragiu, G. Back, The greatest prime factor and related magmas. JP J. Algebra, Number Theory Appl. **15**(2), 127–136 (2009)

M. Caragiu, N. Baxter, A note on *p*-adic Ducci games .JP J. Algebra Number Theory Appl. **8**(1), 115–120 (2007)

M. Caragiu, R. Johns, J. Gieseler, Quasi-random structures from elliptic curves. JP J. Algebra, Number Theory Appl. **6**(3), 561–571 (2006)

M. Caragiu, A. Risch, An Euler-Fibonacci sequence. Far East J. Math. Sci. **52**(1), 1–7 (2011)

M. Caragiu, L. Scheckelhoff, The greatest prime factor and related sequences. JP J. Algebra, Number Theory Appl. **6**(2), 403–409 (2006)

M. Caragiu, P.A. Vicol, Prime magmas and a cyclicity conjecture. JP J. Algebra, Number Theory and Appl. **38**(2), 129–143 (2016)

M. Caragiu, L. Sutherland, M. Zaki, Multidimensional greatest prime factor sequences, JP J. Algebra Number Theory Appl. **23**(2), 187–195 (2011)

M. Caragiu, P.A. Vicol, M. Zaki, On Conway's subprime function, a covering of $\mathbb{N}$ and an unexpected sighting of the Golden Ratio (to appear in Fibonacci Q.)

M. Caragiu, A. Zaharescu, M. Zaki, On Ducci sequences with algebraic numbers. Fibonacci Q. **49** (1), 34–40 (2011)

M. Caragiu, A. Zaharescu, M. Zaki, On a class of solvable recurrences with primes. JP J. Algebra, Number Theory Appl. 26 (2), 197–208 (2012)

M. Caragiu, A. Zaharescu, M. Zaki, An analogue of the Proth-Gilbreath conjecture. Far East J. Math. Sci. 81(1), 1–12 (2013)

M. Caragiu, A. Zaharescu, M. Zaki, On Ducci sequences with primes. Fibonacci Q. **52**(1), 32–38 (2014)

M. Chamberland, Unbounded Ducci sequences. J. Differ. Equ. **9**(10), 887–895 (2003)

M. Chamberland, D.M. Thomas, The N-number Ducci game. J. Differ. Equ. Appl. **10**(3), 339–342 (2004)

Z. Chatzidakis, L. van den Dries, A. Macintyre, Definable sets over finite fields. J. Reine Angew. Math. **427**, 107–135 (1992)

F.R.K. Chung, R.L Graham, R.M.Wilson, Quasi-random graphs. Combinatorica **9**(4), 345–362 (1989)

F.R.K. Chung, R.L Graham, Quasi-random subsets of $\mathbb{Z}_N$. J. Comb. Theory Ser. A **61**(1), 64–86 (1992)

C. Cobeli, A. Zaharescu, A game with divisors and absolute differences of exponents. J. Differ. Equ. Appl. **20**(11) (2014). Also see https://arxiv.org/pdf/1411.1334v1.pdf

H. Cohen, *A Course in Computational Algebraic Number Theory. Graduate Texts in Mathematics* (Springer ,Berlin, 2000)

S.D. Cohen, Consecutive primitive roots in a finite field, in *Proceedings of the American Mathematical Society*, vol. 93, no (2), pp. 189–197 (1985)

G.L. Cohen, P. jun. Hagis, On the number of prime factors of n if $\phi(n)|(n-1)$. Nieuw Arch. Wiskunde, III. Ser. **28**, 177–185 (1980)

S.D. Cohen, G.L. Mullen, Primitive elements in finite fields and costas arrays. Appl. Algebra Eng. Commun. Comput. (2), 45–53 (1991)

J.P. Costas, A study of a class of detection waveforms having nearly ideal range-Doppler ambiguity properties. Proc. IEEE **72**(8), 996–1009 (1984)

L. Curchin, R. Herz-Fischler, De quand date le premier rapprochement entre la suite de Fibonacci et la division en extrême et moyenne raison? Centaurus **28**(2), 129–138 (1985)

A. Das, *Computational Number Theory (Discrete Mathematics and Its Applications)*, 1 edn. (Chapman and Hall/CRC, 2013)

H. Davenport, On the distribution of quadratic residues (mod p). J. London Math. Soc. **6**, 49–54 (1931)

H. Davenport, On the distribution of quadratic residues (mod p), J. London Math .Soc. **8**, 46–52 (1933)

H. Davenport, *Multiplicative Number Theory,* 3rd edn. (Springer, New York,2000)

S. Dehaene, *The Number Sense: How the Mind Creates Mathematics*, first edn. (Oxford University Press, 1999)

E. Delanoy (http://math.stackexchange.com/users/15381/ewan-delanoy), Prove that \$\phi(n) \geq \sqrt{n}/2\$, URL (version: 2013-10-16): http://math.stackexchange.com/q/527966

L.E. Dickson, A new extension of Dirichlet's theorem on prime numbers. Messenger. Math. **33**, 155–161 (1904)

K. Dickman, On the frequency of numbers containing prime factors of a certain relative magnitude. Arkiv för Mat., Astron. och Fys. **22A**, 1–14 (1930)

C. Ding, T. Helleseth, W. Shan, On the linear complexity of legendre sequences. IEEE Trans. Inf. Theory **44**(3), 1276–1278 (1998)

P. Dusart, Autour de la fonction qui compte le nombre de nombres premiers (PhD thesis, in French), http://www.unilim.fr/laco/theses/1998/T1998_01.pdf. Last accessed 14 Oct, 2016

P. Dusart, The k-th prime is greater than $k(\ln k + \ln\ln k - 1)$ for $k \geq 2$. Math. Comput. **68**, 411–415 (1999)

H.M. Edwards, *Riemann's Zeta Function* (Academic Press, 1974)

Euclid Elements (in English and Greek). http://farside.ph.utexas.edu/Books/Euclid/Elements.pdf. Last accessed 13 Oct, 2016

F. d'Errico et al., Early evidence of San material culture represented by organic artifacts from Border Cave, South Africa. PNAS **109**(33), 13214–13219 (2012)

M. Embree, L.N. Trefethen, Growth and decay of random Fibonacci sequences, in *Proceedings of the Royal Society A: Mathematical, Physical and Engineering Sciences,* vol. 455, pp. 2471 (1987)

P. Erdős, On the greatest prime factor of $\prod_{k=1}^{x} f(k)$. J. London Math. Soc. **27**, 379–384 (1952)

P. Erdős, T.N. Shorey, On the greatest prime factor of $2p - 1$ for a prime $p$ and other expressions. Acta Arith. **30**(3), 257–264 (1976)

P. Erdős, J.H. van Lint, On the average ratio of the smallest and largest prime divisor of $n$. Nederl. Akad. Wetensch. Indag. Math. **44**(2), 127–132 (1982)

Experimental Mathematics in Number Theory, Operator Algebras, and Topology, University of Copenhagen. http://www.math.ku.dk/english/research/tfa/xm/. Last accessed 11 Oct, 2016

W.F. Feller, *Introduction to Probability Theory and its Applications*, vol.1, 3rd edition revised printing (1968)

E. Frenkel, *Love and Math: The Heart of Hidden Reality*, 2nd printing edition (Basic Books, 2013)

H. Fukś, N. Boccara, Generalized deterministic traffic rules. Int. J. Modern Phys. C. **9**(1), 1–12 (1998)

M. Fukui, Y. Ishibashi, Traffic flow in 1D cellular automaton model including cars moving with high speed. J. Phys. Soc. Jpn. **65**(6), 1868–1870 (1996)

P.X. Gallagher, On the distribution of primes in short intervals. Mathematika **23**(1), 4–9 (1976)

M. Gardner, Mathematical games—The fantastic combinations of John Conway's new solitaire game `life'. Sci .Am. **223**, 120–123 (1970)

J. von zur Gathen, J. Gerhard, *Modern Computer Algebra*, second edn. (Cambridge University Press, 2003)

Norman Gilbreath - Magicpedia - Genii Magazine, http://www.geniimagazine.com/magicpedia/Norman_Gilbreath

I.M. Gessel, R.P. Stanley, *Algebraic enumeration*, in Handbook of Combinatorics (Vol. II), ed. by R. Graham, M. Grotschel, L. Lovasz first edn. (North Holland, 1995)

K. Gödel, Some basic theorems on the foundations. Collected Works, vol. 3 (unpublished 1951 essay, as cited by Gregory Chaitin in "Two philosophical applications of algorithmic information theory"—ArXiV https://arxiv.org/pdf/math/0302333.pdf—there Chaitin argues that despite Gödel's platonism, he actually manages to arrive, at least partially, to a "pseudo-empirical or a quasi-empirical position")

Christian Goldbach's letter to Euler, http://eulerarchive.maa.org//correspondence/letters/OO0765.pdf. Last accessed on 16 Oct, 2016

T. Gowers, *Mathematics: A Very Short Introduction* (Oxford University Press, 2002)

A. Granville, G. Martin, Prime number races. Am. Math. Mon. **113**(1), 1–33 (2006)

C.R. Greathouse IV, Historical sequences. From the On-Line Encyclopedia of Integer Sequences® (OEIS®) wiki. Available at https://oeis.org/wiki/Historical_sequences

B. Green, T. Tao, The primes contain arbitrarily long arithmetic progressions. Ann. Math. **167**(2), 481–547 (2008)

B. Green, T. Tao, Linear equations in primes. Ann. Math. **171**(2010), 1753–1850

E. Grosswald, On Burgess' bound for primitive roots modulo primes and an application to $\Gamma(p)$ Am. J. Math. **103**(6), 1171–1183 (1981)

A. Grytczuk, M. Wójtowicz, F. Luca, Another note on the greatest prime factors of Fermat numbers. Southeast Asian Bull. Math. **25**(1), 111–115 (2001)

R.K. Guy, T. Khovanova, J. Salazar, Conway's subprime Fibonacci sequences. Math. Mag. **87**(5), 323–337 (December 2014)

G.H. Hardy, E.M. Wright, *An Introduction to the Theory of Numbers*, 5th edn. (Clarendon Press, Oxford, England, 1979)

M. Haver, On the R. Lemke Oliver—K. Soundararajan recent "prime conspiracy". Talk delivered at the Spring 2016 Ohio MAA Meeting. Ohio Norther University, April 8, 2016.

M. Haver, Poissonian character and Chebyshev bias for greatest prime factor sequences: a computational analysis. Senior Capstone Project. Ohio Northern University, May 2017 (advisor M. Caragiu)

C.S. Heyde, On a probabilistic analogue of the Fibonacci sequence. J. Appl. Prob. **17**, 1079–1082 (1980)

T. Hill, A statistical derivation of the significant-digit law. Statist. Sci. **10**, 354–363 (1996)

R. Honsberger, *Ingenuity in Mathematics* (Yale University, New York, 1970), pp. 80–83

C. Hooley, On the greatest prime factor of a quadratic polynomial. Acta Math. 117, 281–299 (1967)

A. Ilachinski, *Cellular Automata: A Discrete Universe*, Reprint Edition. (World Scientific, 2001)

J.G. Kemeny, Largest prime factor. J. Pure Appl. Algebra **89**(1–2), 181–186 (1993)

J. Kepler, The six-cornered snowflake (1611). Translated by Colin Hardie (Clarendon Press, Oxford, 1966)

T. Khovanova, Conway's Subprime Fibonacci Sequences, Tatiana Khovanova blog, 30th July 2012 entry, blog.tanyakhovanova.com/2012/07/conways-subprime-fibonacci-sequences/

E. Klarreich, Mathematicians discover prime conspiracy. Quanta Mag. (March 13, 2016)

S. Knapowski, P. Turan, Comparative prime-number theory, I-III. Acta Math. Acad. Sci. Hungar. **13**, 299–364 (1962)

R. Knott, Fibonacci and Golden Ratio Formulae. http://www.maths.surrey.ac.uk/hosted-sites/R. Knott/Fibonacci/fibFormulae.html. Updated 25 September 2016

I. Krasikov, J.C. Lagarias, Bounds for the 3x+1 problem using difference inequalities. Acta Arith. **109**(3), 237–258 (2003)

D. Knuth, *Semi Numerical Algorithms*, Second Edition. (Addison-Wesley, 1969)

D.E. Knuth, L.T. Pardo, Analysis of a simple factorization algorithm. Theor. Comput. Sci. **3**, 321–348 (1976)

A.V. Kontorovich, S.J. Miller, Benford's law, values of L-functions and the 3x+1 problem. Acta Arith. **120**(3), 269–297 (2005)

G.D. Kuh, *High-Impact Educational Practices: What They Are, Who Has Access to Them, and Why They Matter* (AAC&U, 2008)

J.C. Lagarias, The 3x+ 1 problem and its generalizations. Am. Math. Mon. **92**, 3–23 (1985)

J.C. Lagarias, K. Soundararajan, Benford's Law for the 3x+1 function. J. London Math. Soc. **74** (2), 289-303 (2006)

S. Laishram, T.N. Shorey, The greatest prime divisor of a product of consecutive integers. Acta Arith. **120**(3), 299–306 (2005)

R.J. Lemke Oliver, K. Soundararajan, Unexpected biases in the distribution of consecutive primes. E-print arXiv:1603.03720 (March 11, 2016)

W.J. Leveque, *Fundamentals of Number Theory* (Dover, 1977)

J.E. Littlewood, Distribution des nombres premiers. C.R. Acad. Sci. Paris **158**, 1869–1872 (1914)

M. Livio, *The Golden Ratio: The Story of Phi, the World's Most Astonishing Number* (Broadway Books, New York, 2002)

F. Luca, F. Najman, On the largest prime factor of $x^2 - 1$. Math. Comp. **80**(273), 429–435 (2011)

F. Luca, Arithmetic functions of Fibonacci numbers. Fibonacci Q. **37**(3), 265–268 (1999)

F. Luca, Equations involving arithmetic functions of Fibonacci and Lucas numbers. Fibonacci Q. **38**(1), 49–55 (2000)

F. Luca, V.J. Meja Huguet, F. Nicolae, On the Euler function of Fibonacci numbers. J. Integer Seq. **12**(6), Article 09.6.6 (2009)

F. Luca, P. Stănică, Linear equations with the Euler's totient function. Acta Arith. **128**(2), 135-147 (2007)

F.L. Mannering, W.P. Kilareski, *Principles of Highway Engineering and Traffic Analysis*. (Wiley, 1990)

J.L. Massey, Shift-register synthesis and BCH decoding. IEEE Trans. Inf. Theory, IT-15 (1), 122–127 (1969)

O. Martin, A.M. Odlyzko, S. Wolfram, Algebraic properties of cellular automata. Commun. Math. Phys. **93**(2), 219–258 (1984)

A.D. May, *Traffic Flow Fundamentals* (Prentice Hall, 1990)

K. McSpadden, You Now have a shorter attention span than a goldfish. TIME Mag. (May 14, 2015). http://time.com/3858309/attention-spans-goldfish. Accessed on 5 Oct, 2016

P.E. Mendelsohn, The Pentanacci numbers. The Fibonacci sequence. A Collection of manuscripts related to the Fibonacci sequence: 18th anniversary volume (1980), pp. 31–33

I.D. Mercer, Autocorrelations of random binary sequences. Comb. Probab. Comput. **15**(5), 663–671 (2006)

M. Monagan, Computations on the 3n + 1 conjecture. Worksheet created by Michael Monagan, MAPLE application www.maplesoft.com/applications/view.aspx?SID=3644&view=html

O. Moreno, J. Sotero, Computational approach to "Conjecture A" of Golomb, Congr. Numer. **70**, 216 (1990)

J. Nagura, On the interval containing at least one prime number in *Proceedings of the Japan Academy, Series A,* vol. 28 (1952), pp. 177–181.

T. Noe, T. Piezas III, E.W. Weisstein, Fibonacci n-Step Number. From MathWorld—A Wolfram Web Resource. http://mathworld.wolfram.com/Fibonaccin-StepNumber.html

A.M. Odlyzko, Iterated absolute values of differences of consecutive primes. Math. Comput. **61**, 373–380 (1993)

The Online Encyclopedia of Integer Sequences. https://oeis.org/

Nobelprize.org. Press Release: The Nobel Prize in Physics 2016. https://www.nobelprize.org/nobel_prizes/physics/laureates/2016/press.html. Accessed on 5 Oct, 2016

R. Peralta, On the distribution of quadratic residues and nonresidues modulo a prime number. Math. Comput. **58**(197), 433-440 (1992)

J.M. Pollard, A Monte Carlo method for factorization. BIT Numer. Math. **15**(3), 331–334 (1975)

H. Qi, Stream Ciphers and Linear Complexity. Masters' Thesis. University of Maryland (2007)

M. Renault, The Fibonacci sequence under various moduli. M.Sc. Thesis. Wake Forest University (1996). http://webspace.ship.edu/msrenault/fibonacci/FibThesis.pdf

D.A. Rosenblueth, C. Gershenson, A model of city traffic based on elementary cellular automata. Complex Syst. **19**(4) (2011)

J.B. Rosser, L. Schoenfeld, Approximate formulas for some functions of prime numbers. Ill. J. Math. **6**, 64–94 (1962)

S. Roberts, Sciences live: John Conway. Simon Foundation Blog entry—April 4, 2014. https://www.simonsfoundation.org/science_lives_video/john-conway/

S. Roberts, Genius at play: the curious mathematical mind of John Horton Conway. Bloomsbury, reprint edition (2016)

A. Schinzel, W. Sierpinski, Sur certaines hypothèses concernant les nombres premiers, Acta Arith. IV (1958) 185–208; corrigé ibid. V (1958) 259.

J.C. Schroeder, Small special pairs of primitive roots. Senior Capstone Project, Ohio Northern University, December 2013 (advisor M. Caragiu)

V. Shoup, *A Computational Introduction to Number Theory and Algebra*, Second edition. (Cambridge University Press, 2009)

L. Pisano, Fibonacci's Liber Abaci: A Translation into Modern English of the Book of Calculation. Sources and Studies in the History of Mathematics and Physical Sciences, Sigler, Laurence E, trans. (Springer, Berlin, 2002)

P. Singh, The So-called Fibonacci numbers in ancient and medieval India. Hist. Math. **12**(3), 229–44 (1985)

N.J.A. Sloane, The online encyclopedia of integer sequences. Ann. Math. Inf. **41**, 219–234 (2013)

L. Sutherland, Multidimensional greatest prime factor sequences, senior capstone project, Ohio Northern University. Advisor: M. Caragiu. (2011)

The Mathematical Tourist blog. Wild Beasts around the Corner. March 17, 2013. http://mathtourist.blogspot.ca/2013/03/wild-beasts-around-corner.html

E. Vegh, Pairs of consecutive primitive roots modulo a prime. Proc. Am. Math. Soc. **19**, 1169–1170 (1968)

D. Viswanath, Random Fibonacci sequences and the number 1.13198824... Math. Comput. **69** (231), 1131–1155 (1999)

J. Wang, X. Wang, On the set of reduced $\phi$-partitions of a positive integer. Fibonacci Q. **44**(2), 98–102 (2006)

W.A. Webb, The length of the four-number game. Fibonacci Q. **20**(1), 33–35. http://www.fq.math.ca/Scanned/20-1/webb.pdf

A. Weil, Numbers of solutions of equations in finite fields. Bull. Am. Math. Soc. 55, 497–508 (1949)

L. Wentian, Power spectra of regular languages and cellular automata. Complex Syst. 1, 107–130 (1987)

S. Wolfram, Statistical mechanics of cellular automata. Rev. Modern Phys. **55**(3) (July 1983)

E.W. Weisstein, Rule 102. From MathWorld—A Wolfram Web Resource. http://mathworld. wolfram.com/Rule102.html

E.W. Weisstein, Cellular Automaton. From MathWorld—A Wolfram Web Resource. http:// mathworld.wolfram.com/CellularAutomaton.html

E.W. Weisstein, Moore Neighborhood. From MathWorld—A Wolfram Web Resource. http:// mathworld.wolfram.com/MooreNeighborhood.html

E.W. Weisstein, Random Walk—1-Dimensional. From MathWorld—A Wolfram Web Resource. http://mathworld.wolfram.com/RandomWalk1-Dimensional.html

E.W. Weisstein, Golomb-Dickman Constant. From MathWorld—A Wolfram Web Resource. http://mathworld.wolfram.com/Golomb-DickmanConstant.html

Printed in the United States
By Bookmasters